ISBN 0-8373-2912-4

C-2912 CAREER EXAMINATION SERIES

This is your PASSBOOK® for...

Mosquito Control Inspector

Test Preparation Study Guide

Questions & Answers

NLC

NATIONAL LEARNING CORPORATION

National Learning Corporation

212 Michael Drive, Syosset, NY 11791
(516) 921-8888 • www.passbooks.com
E-mail: info@passbooks.com

PUBLISHED IN THE UNITED STATES OF AMERICA

PASSBOOK® SERIES

THE *PASSBOOK® SERIES* has been created to prepare applicants and candidates for the ultimate academic battlefield – the examination room.

At some time in our lives, each and every one of us may be required to take an examination – for validation, matriculation, admission, qualification, registration, certification, or licensure.

Based on the assumption that every applicant or candidate has met the basic formal educational standards, has taken the required number of courses, and read the necessary texts, the *PASSBOOK® SERIES* furnishes the one special preparation which may assure passing with confidence, instead of failing with insecurity. Examination questions – together with answers – are furnished as the basic vehicle for study so that the mysteries of the examination and its compounding difficulties may be eliminated or diminished by a sure method.

This book is meant to help you pass your examination provided that you qualify and are serious in your objective.

The entire field is reviewed through the huge store of content information which is succinctly presented through a provocative and challenging approach – the question-and-answer method.

A climate of success is established by furnishing the correct answers at the end of each test.

You soon learn to recognize types of questions, forms of questions, and patterns of questioning. You may even begin to anticipate expected outcomes.

You perceive that many questions are repeated or adapted so that you can gain acute insights, which may enable you to score many sure points.

You learn how to confront new questions, or types of questions, and to attack them confidently and work out the correct answers.

You note objectives and emphases, and recognize pitfalls and dangers, so that you may make positive educational adjustments.

Moreover, you are kept fully informed in relation to new concepts, methods, practices, and directions in the field.

You discover that you are actually taking the examination all the time: you are preparing for the examination by "taking" an examination, not by reading extraneous and/or supererogatory textbooks.

In short, this PASSBOOK®, used directedly, should be an important factor in helping you to pass your test.

MOSQUITO CONTROL INSPECTOR

DUTIES

Supervises extermination and control of mosquitoes in an assigned area; performs related duties as required.

SUBJECT OF EXAMINATION

Written test designed to test for knowledge, skills, and/or abilities in such areas as:

1. Mosquito and insect control and extermination;
2. Keeping simple inventory records;
3. Understanding an interpreting written material;
4. Organizing data into tables and records; and
5. Supervision.

HOW TO TAKE A TEST

I. YOU MUST PASS AN EXAMINATION

A. WHAT EVERY CANDIDATE SHOULD KNOW

Examination applicants often ask us for help in preparing for the written test. What can I study in advance? What kinds of questions will be asked? How will the test be given? How will the papers be graded?

As an applicant for a civil service examination, you may be wondering about some of these things. Our purpose here is to suggest effective methods of advance study and to describe civil service examinations.

Your chances for success on this examination can be increased if you know how to prepare. Those "pre-examination jitters" can be reduced if you know what to expect. You can even experience an adventure in good citizenship if you know why civil service exams are given.

B. WHY ARE CIVIL SERVICE EXAMINATIONS GIVEN?

Civil service examinations are important to you in two ways. As a citizen, you want public jobs filled by employees who know how to do their work. As a job seeker, you want a fair chance to compete for that job on an equal footing with other candidates. The best-known means of accomplishing this two-fold goal is the competitive examination.

Exams are widely publicized throughout the nation. They may be administered for jobs in federal, state, city, municipal, town or village governments or agencies.

Any citizen may apply, with some limitations, such as the age or residence of applicants. Your experience and education may be reviewed to see whether you meet the requirements for the particular examination. When these requirements exist, they are reasonable and applied consistently to all applicants. Thus, a competitive examination may cause you some uneasiness now, but it is your privilege and safeguard.

C. HOW ARE CIVIL SERVICE EXAMS DEVELOPED?

Examinations are carefully written by trained technicians who are specialists in the field known as "psychological measurement," in consultation with recognized authorities in the field of work that the test will cover. These experts recommend the subject matter areas or skills to be tested; only those knowledges or skills important to your success on the job are included. The most reliable books and source materials available are used as references. Together, the experts and technicians judge the difficulty level of the questions.

Test technicians know how to phrase questions so that the problem is clearly stated. Their ethics do not permit "trick" or "catch" questions. Questions may have been tried out on sample groups, or subjected to statistical analysis, to determine their usefulness.

Written tests are often used in combination with performance tests, ratings of training and experience, and oral interviews. All of these measures combine to form the best-known means of finding the right person for the right job.

II. HOW TO PASS THE WRITTEN TEST

A. NATURE OF THE EXAMINATION

To prepare intelligently for civil service examinations, you should know how they differ from school examinations you have taken. In school you were assigned certain definite pages to read or subjects to cover. The examination questions were quite detailed and usually emphasized memory. Civil service exams, on the other hand, try to discover your present ability to perform the duties of a position, plus your potentiality to learn these duties. In other words, a civil service exam attempts to predict how successful you will be. Questions cover such a broad area that they cannot be as minute and detailed as school exam questions.

In the public service similar kinds of work, or positions, are grouped together in one "class." This process is known as *position-classification.* All the positions in a class are paid according to the salary range for that class. One class title covers all of these positions, and they are all tested by the same examination.

B. FOUR BASIC STEPS

1) Study the announcement

How, then, can you know what subjects to study? Our best answer is: "Learn as much as possible about the class of positions for which you've applied." The exam will test the knowledge, skills and abilities needed to do the work.

Your most valuable source of information about the position you want is the official exam announcement. This announcement lists the training and experience qualifications. Check these standards and apply only if you come reasonably close to meeting them.

The brief description of the position in the examination announcement offers some clues to the subjects which will be tested. Think about the job itself. Review the duties in your mind. Can you perform them, or are there some in which you are rusty? Fill in the blank spots in your preparation.

Many jurisdictions preview the written test in the exam announcement by including a section called "Knowledge and Abilities Required," "Scope of the Examination," or some similar heading. Here you will find out specifically what fields will be tested.

2) Review your own background

Once you learn in general what the position is all about, and what you need to know to do the work, ask yourself which subjects you already know fairly well and which need improvement. You may wonder whether to concentrate on improving your strong areas or on building some background in your fields of weakness. When the announcement has specified "some knowledge" or "considerable knowledge," or has used adjectives like "beginning principles of..." or "advanced ... methods," you can get a clue as to the number and difficulty of questions to be asked in any given field. More questions, and hence broader coverage, would be included for those subjects which are more important in the work. Now weigh your strengths and weaknesses against the job requirements and prepare accordingly.

3) Determine the level of the position

Another way to tell how intensively you should prepare is to understand the level of the job for which you are applying. Is it the entering level? In other words, is this the position in which beginners in a field of work are hired? Or is it an intermediate or advanced level? Sometimes this is indicated by such words as "Junior" or "Senior" in the class title. Other jurisdictions use Roman numerals to designate the level – Clerk I, Clerk II, for example. The word "Supervisor" sometimes appears in the title. If the level is not indicated by the title, check the description of duties. Will you be working under very close supervision, or will you have responsibility for independent decisions in this work?

4) Choose appropriate study materials

Now that you know the subjects to be examined and the relative amount of each subject to be covered, you can choose suitable study materials. For beginning level jobs, or even advanced ones, if you have a pronounced weakness in some aspect of your training, read a modern, standard textbook in that field. Be sure it is up to date and has general coverage. Such books are normally available at your library, and the librarian will be glad to help you locate one. For entry-level positions, questions of appropriate difficulty are chosen – neither highly advanced questions, nor those too simple. Such questions require careful thought but not advanced training.

If the position for which you are applying is technical or advanced, you will read more advanced, specialized material. If you are already familiar with the basic principles of your field, elementary textbooks would waste your time. Concentrate on advanced textbooks and technical periodicals. Think through the concepts and review difficult problems in your field.

These are all general sources. You can get more ideas on your own initiative, following these leads. For example, training manuals and publications of the government agency which employs workers in your field can be useful, particularly for technical and professional positions. A letter or visit to the government department involved may result in more specific study suggestions, and certainly will provide you with a more definite idea of the exact nature of the position you are seeking.

III. KINDS OF TESTS

Tests are used for purposes other than measuring knowledge and ability to perform specified duties. For some positions, it is equally important to test ability to make adjustments to new situations or to profit from training. In others, basic mental abilities not dependent on information are essential. Questions which test these things may not appear as pertinent to the duties of the position as those which test for knowledge and information. Yet they are often highly important parts of a fair examination. For very general questions, it is almost impossible to help you direct your study efforts. What we can do is to point out some of the more common of these general abilities needed in public service positions and describe some typical questions.

1) General information

Broad, general information has been found useful for predicting job success in some kinds of work. This is tested in a variety of ways, from vocabulary lists to questions about current events. Basic background in some field of work, such as sociology or economics, may be sampled in a group of questions. Often these are principles which have become familiar to most persons through exposure rather than through formal training. It is difficult to advise you how to study for these questions; being alert to the world around you is our best suggestion.

2) Verbal ability

An example of an ability needed in many positions is verbal or language ability. Verbal ability is, in brief, the ability to use and understand words. Vocabulary and grammar tests are typical measures of this ability. Reading comprehension or paragraph interpretation questions are common in many kinds of civil service tests. You are given a paragraph of written material and asked to find its central meaning.

3) Numerical ability

Number skills can be tested by the familiar arithmetic problem, by checking paired lists of numbers to see which are alike and which are different, or by interpreting charts and graphs. In the latter test, a graph may be printed in the test booklet which you are asked to use as the basis for answering questions.

4) Observation

A popular test for law-enforcement positions is the observation test. A picture is shown to you for several minutes, then taken away. Questions about the picture test your ability to observe both details and larger elements.

5) Following directions

In many positions in the public service, the employee must be able to carry out written instructions dependably and accurately. You may be given a chart with several columns, each column listing a variety of information. The questions require you to carry out directions involving the information given in the chart.

6) Skills and aptitudes

Performance tests effectively measure some manual skills and aptitudes. When the skill is one in which you are trained, such as typing or shorthand, you can practice. These tests are often very much like those given in business school or high school courses. For many of the other skills and aptitudes, however, no short-time preparation can be made. Skills and abilities natural to you or that you have developed throughout your lifetime are being tested.

Many of the general questions just described provide all the data needed to answer the questions and ask you to use your reasoning ability to find the answers. Your best preparation for these tests, as well as for tests of facts and ideas, is to be at your physical and mental best. You, no doubt, have your own methods of getting into an exam-taking mood and keeping "in shape." The next section lists some ideas on this subject.

IV. KINDS OF QUESTIONS

Only rarely is the "essay" question, which you answer in narrative form, used in civil service tests. Civil service tests are usually of the short-answer type. Full instructions for answering these questions will be given to you at the examination. But in case this is your first experience with short-answer questions and separate answer sheets, here is what you need to know:

1) Multiple-choice Questions

Most popular of the short-answer questions is the "multiple choice" or "best answer" question. It can be used, for example, to test for factual knowledge, ability to solve problems or judgment in meeting situations found at work.

A multiple-choice question is normally one of three types—

- It can begin with an incomplete statement followed by several possible endings. You are to find the one ending which *best* completes the statement, although some of the others may not be entirely wrong.
- It can also be a complete statement in the form of a question which is answered by choosing one of the statements listed.

- It can be in the form of a problem – again you select the best answer.

Here is an example of a multiple-choice question with a discussion which should give you some clues as to the method for choosing the right answer:

When an employee has a complaint about his assignment, the action which will *best* help him overcome his difficulty is to

A. discuss his difficulty with his coworkers
B. take the problem to the head of the organization
C. take the problem to the person who gave him the assignment
D. say nothing to anyone about his complaint

In answering this question, you should study each of the choices to find which is best. Consider choice "A" – Certainly an employee may discuss his complaint with fellow employees, but no change or improvement can result, and the complaint remains unresolved. Choice "B" is a poor choice since the head of the organization probably does not know what assignment you have been given, and taking your problem to him is known as "going over the head" of the supervisor. The supervisor, or person who made the assignment, is the person who can clarify it or correct any injustice. Choice "C" is, therefore, correct. To say nothing, as in choice "D," is unwise. Supervisors have and interest in knowing the problems employees are facing, and the employee is seeking a solution to his problem.

2) True/False Questions

The "true/false" or "right/wrong" form of question is sometimes used. Here a complete statement is given. Your job is to decide whether the statement is right or wrong.

SAMPLE: A roaming cell-phone call to a nearby city costs less than a non-roaming call to a distant city.

This statement is wrong, or false, since roaming calls are more expensive.

This is not a complete list of all possible question forms, although most of the others are variations of these common types. You will always get complete directions for answering questions. Be sure you understand *how* to mark your answers – ask questions until you do.

V. RECORDING YOUR ANSWERS

Computer terminals are used more and more today for many different kinds of exams.

For an examination with very few applicants, you may be told to record your answers in the test booklet itself. Separate answer sheets are much more common. If this separate answer sheet is to be scored by machine – and this is often the case – it is highly important that you mark your answers correctly in order to get credit.

An electronic scoring machine is often used in civil service offices because of the speed with which papers can be scored. Machine-scored answer sheets must be marked with a pencil, which will be given to you. This pencil has a high graphite content which responds to the electronic scoring machine. As a matter of fact, stray dots may register as answers, so do not let your pencil rest on the answer sheet while you are pondering the correct answer. Also, if your pencil lead breaks or is otherwise defective, ask for another.

Since the answer sheet will be dropped in a slot in the scoring machine, be careful not to bend the corners or get the paper crumpled.

The answer sheet normally has five vertical columns of numbers, with 30 numbers to a column. These numbers correspond to the question numbers in your test booklet. After each number, going across the page are four or five pairs of dotted lines. These short dotted lines have small letters or numbers above them. The first two pairs may also have a "T" or "F" above the letters. This indicates that the first two pairs only are to be used if the questions are of the true-false type. If the questions are multiple choice, disregard the "T" and "F" and pay attention only to the small letters or numbers.

Answer your questions in the manner of the sample that follows:

32. The largest city in the United States is
    A. Washington, D.C.
    B. New York City
    C. Chicago
    D. Detroit
    E. San Francisco

1) Choose the answer you think is best. (New York City is the largest, so "B" is correct.)
2) Find the row of dotted lines numbered the same as the question you are answering. (Find row number 32)
3) Find the pair of dotted lines corresponding to the answer. (Find the pair of lines under the mark "B.")
4) Make a solid black mark between the dotted lines.

VI. BEFORE THE TEST

Common sense will help you find procedures to follow to get ready for an examination. Too many of us, however, overlook these sensible measures. Indeed, nervousness and fatigue have been found to be the most serious reasons why applicants fail to do their best on civil service tests. Here is a list of reminders:

- Begin your preparation early – Don't wait until the last minute to go scurrying around for books and materials or to find out what the position is all about.
- Prepare continuously – An hour a night for a week is better than an all-night cram session. This has been definitely established. What is more, a night a week for a month will return better dividends than crowding your study into a shorter period of time.
- Locate the place of the exam – You have been sent a notice telling you when and where to report for the examination. If the location is in a different town or otherwise unfamiliar to you, it would be well to inquire the best route and learn something about the building.
- Relax the night before the test – Allow your mind to rest. Do not study at all that night. Plan some mild recreation or diversion; then go to bed early and get a good night's sleep.
- Get up early enough to make a leisurely trip to the place for the test – This way unforeseen events, traffic snarls, unfamiliar buildings, etc. will not upset you.
- Dress comfortably – A written test is not a fashion show. You will be known by number and not by name, so wear something comfortable.

- Leave excess paraphernalia at home – Shopping bags and odd bundles will get in your way. You need bring only the items mentioned in the official notice you received; usually everything you need is provided. Do not bring reference books to the exam. They will only confuse those last minutes and be taken away from you when in the test room.
- Arrive somewhat ahead of time – If because of transportation schedules you must get there very early, bring a newspaper or magazine to take your mind off yourself while waiting.
- Locate the examination room – When you have found the proper room, you will be directed to the seat or part of the room where you will sit. Sometimes you are given a sheet of instructions to read while you are waiting. Do not fill out any forms until you are told to do so; just read them and be prepared.
- Relax and prepare to listen to the instructions
- If you have any physical problem that may keep you from doing your best, be sure to tell the test administrator. If you are sick or in poor health, you really cannot do your best on the exam. You can come back and take the test some other time.

VII. AT THE TEST

The day of the test is here and you have the test booklet in your hand. The temptation to get going is very strong. Caution! There is more to success than knowing the right answers. You must know how to identify your papers and understand variations in the type of short-answer question used in this particular examination. Follow these suggestions for maximum results from your efforts:

1) Cooperate with the monitor

The test administrator has a duty to create a situation in which you can be as much at ease as possible. He will give instructions, tell you when to begin, check to see that you are marking your answer sheet correctly, and so on. He is not there to guard you, although he will see that your competitors do not take unfair advantage. He wants to help you do your best.

2) Listen to all instructions

Don't jump the gun! Wait until you understand all directions. In most civil service tests you get more time than you need to answer the questions. So don't be in a hurry. Read each word of instructions until you clearly understand the meaning. Study the examples, listen to all announcements and follow directions. Ask questions if you do not understand what to do.

3) Identify your papers

Civil service exams are usually identified by number only. You will be assigned a number; you must not put your name on your test papers. Be sure to copy your number correctly. Since more than one exam may be given, copy your exact examination title.

4) Plan your time

Unless you are told that a test is a "speed" or "rate of work" test, speed itself is usually not important. Time enough to answer all the questions will be provided, but this does not mean that you have all day. An overall time limit has been set. Divide the total time (in minutes) by the number of questions to determine the approximate time you have for each question.

5) Do not linger over difficult questions

If you come across a difficult question, mark it with a paper clip (useful to have along) and come back to it when you have been through the booklet. One caution if you do this – be sure to skip a number on your answer sheet as well. Check often to be sure that you have not lost your place and that you are marking in the row numbered the same as the question you are answering.

6) Read the questions

Be sure you know what the question asks! Many capable people are unsuccessful because they failed to *read* the questions correctly.

7) Answer all questions

Unless you have been instructed that a penalty will be deducted for incorrect answers, it is better to guess than to omit a question.

8) Speed tests

It is often better NOT to guess on speed tests. It has been found that on timed tests people are tempted to spend the last few seconds before time is called in marking answers at random – without even reading them – in the hope of picking up a few extra points. To discourage this practice, the instructions may warn you that your score will be "corrected" for guessing. That is, a penalty will be applied. The incorrect answers will be deducted from the correct ones, or some other penalty formula will be used.

9) Review your answers

If you finish before time is called, go back to the questions you guessed or omitted to give them further thought. Review other answers if you have time.

10) Return your test materials

If you are ready to leave before others have finished or time is called, take ALL your materials to the monitor and leave quietly. Never take any test material with you. The monitor can discover whose papers are not complete, and taking a test booklet may be grounds for disqualification.

VIII. EXAMINATION TECHNIQUES

1) Read the general instructions carefully. These are usually printed on the first page of the exam booklet. As a rule, these instructions refer to the timing of the examination; the fact that you should not start work until the signal and must stop work at a signal, etc. If there are any *special* instructions, such as a choice of questions to be answered, make sure that you note this instruction carefully.

2) When you are ready to start work on the examination, that is as soon as the signal has been given, read the instructions to each question booklet, underline any key words or phrases, such as *least*, *best*, *outline*, *describe* and the like. In this way you will tend to answer as requested rather than discover on reviewing your paper that you *listed without describing*, that you selected the *worst* choice rather than the *best* choice, etc.

3) If the examination is of the objective or multiple-choice type – that is, each question will also give a series of possible answers: A, B, C or D, and you are called upon to select the best answer and write the letter next to that answer on your answer paper – it is advisable to start answering each question in turn. There may be anywhere from 50 to 100 such questions in the three or four hours allotted and you can see how much time would be taken if you read through all the questions before beginning to answer any. Furthermore, if you come across a question or group of questions which you know would be difficult to answer, it would undoubtedly affect your handling of all the other questions.

4) If the examination is of the essay type and contains but a few questions, it is a moot point as to whether you should read all the questions before starting to answer any one. Of course, if you are given a choice – say five out of seven and the like – then it is essential to read all the questions so you can eliminate the two that are most difficult. If, however, you are asked to answer all the questions, there may be danger in trying to answer the easiest one first because you may find that you will spend too much time on it. The best technique is to answer the first question, then proceed to the second, etc.

5) Time your answers. Before the exam begins, write down the time it started, then add the time allowed for the examination and write down the time it must be completed, then divide the time available somewhat as follows:
   - If 3-1/2 hours are allowed, that would be 210 minutes. If you have 80 objective-type questions, that would be an average of 2-1/2 minutes per question. Allow yourself no more than 2 minutes per question, or a total of 160 minutes, which will permit about 50 minutes to review.
   - If for the time allotment of 210 minutes there are 7 essay questions to answer, that would average about 30 minutes a question. Give yourself only 25 minutes per question so that you have about 35 minutes to review.

6) The most important instruction is to *read each question* and make sure you know what is wanted. The second most important instruction is to *time yourself properly* so that you answer every question. The third most important instruction is to *answer every question*. Guess if you have to but include something for each question. Remember that you will receive no credit for a blank and will probably receive some credit if you write something in answer to an essay question. If you guess a letter – say "B" for a multiple-choice question – you may have guessed right. If you leave a blank as an answer to a multiple-choice question, the examiners may respect your feelings but it will not add a point to your score. Some exams may penalize you for wrong answers, so in such cases *only*, you may not want to guess unless you have some basis for your answer.

7) Suggestions
   a. Objective-type questions
      1. Examine the question booklet for proper sequence of pages and questions
      2. Read all instructions carefully
      3. Skip any question which seems too difficult; return to it after all other questions have been answered
      4. Apportion your time properly; do not spend too much time on any single question or group of questions

5. Note and underline key words – *all, most, fewest, least, best, worst, same, opposite,* etc.
6. Pay particular attention to negatives
7. Note unusual option, e.g., unduly long, short, complex, different or similar in content to the body of the question
8. Observe the use of "hedging" words – *probably, may, most likely,* etc.
9. Make sure that your answer is put next to the same number as the question
10. Do not second-guess unless you have good reason to believe the second answer is definitely more correct
11. Cross out original answer if you decide another answer is more accurate; do not erase until you are ready to hand your paper in
12. Answer all questions; guess unless instructed otherwise
13. Leave time for review

b. Essay questions
1. Read each question carefully
2. Determine exactly what is wanted. Underline key words or phrases.
3. Decide on outline or paragraph answer
4. Include many different points and elements unless asked to develop any one or two points or elements
5. Show impartiality by giving pros and cons unless directed to select one side only
6. Make and write down any assumptions you find necessary to answer the questions
7. Watch your English, grammar, punctuation and choice of words
8. Time your answers; don't crowd material

8) Answering the essay question

Most essay questions can be answered by framing the specific response around several key words or ideas. Here are a few such key words or ideas:

M's: manpower, materials, methods, money, management
P's: purpose, program, policy, plan, procedure, practice, problems, pitfalls, personnel, public relations

a. Six basic steps in handling problems:
1. Preliminary plan and background development
2. Collect information, data and facts
3. Analyze and interpret information, data and facts
4. Analyze and develop solutions as well as make recommendations
5. Prepare report and sell recommendations
6. Install recommendations and follow up effectiveness

b. Pitfalls to avoid
1. *Taking things for granted* – A statement of the situation does not necessarily imply that each of the elements is necessarily true; for example, a complaint may be invalid and biased so that all that can be taken for granted is that a complaint has been registered

2. *Considering only one side of a situation* – Wherever possible, indicate several alternatives and then point out the reasons you selected the best one
3. *Failing to indicate follow up* – Whenever your answer indicates action on your part, make certain that you will take proper follow-up action to see how successful your recommendations, procedures or actions turn out to be
4. *Taking too long in answering any single question* – Remember to time your answers properly

IX. AFTER THE TEST

Scoring procedures differ in detail among civil service jurisdictions although the general principles are the same. Whether the papers are hand-scored or graded by machine we have described, they are nearly always graded by number. That is, the person who marks the paper knows only the number – never the name – of the applicant. Not until all the papers have been graded will they be matched with names. If other tests, such as training and experience or oral interview ratings have been given, scores will be combined. Different parts of the examination usually have different weights. For example, the written test might count 60 percent of the final grade, and a rating of training and experience 40 percent. In many jurisdictions, veterans will have a certain number of points added to their grades.

After the final grade has been determined, the names are placed in grade order and an eligible list is established. There are various methods for resolving ties between those who get the same final grade – probably the most common is to place first the name of the person whose application was received first. Job offers are made from the eligible list in the order the names appear on it. You will be notified of your grade and your rank as soon as all these computations have been made. This will be done as rapidly as possible.

People who are found to meet the requirements in the announcement are called "eligibles." Their names are put on a list of eligible candidates. An eligible's chances of getting a job depend on how high he stands on this list and how fast agencies are filling jobs from the list.

When a job is to be filled from a list of eligibles, the agency asks for the names of people on the list of eligibles for that job. When the civil service commission receives this request, it sends to the agency the names of the three people highest on this list. Or, if the job to be filled has specialized requirements, the office sends the agency the names of the top three persons who meet these requirements from the general list.

The appointing officer makes a choice from among the three people whose names were sent to him. If the selected person accepts the appointment, the names of the others are put back on the list to be considered for future openings.

That is the rule in hiring from all kinds of eligible lists, whether they are for typist, carpenter, chemist, or something else. For every vacancy, the appointing officer has his choice of any one of the top three eligibles on the list. This explains why the person whose name is on top of the list sometimes does not get an appointment when some of the persons lower on the list do. If the appointing officer chooses the second or third eligible, the No. 1 eligible does not get a job at once, but stays on the list until he is appointed or the list is terminated.

X. HOW TO PASS THE INTERVIEW TEST

The examination for which you applied requires an oral interview test. You have already taken the written test and you are now being called for the interview test – the final part of the formal examination.

You may think that it is not possible to prepare for an interview test and that there are no procedures to follow during an interview. Our purpose is to point out some things you can do in advance that will help you and some good rules to follow and pitfalls to avoid while you are being interviewed.

What is an interview supposed to test?

The written examination is designed to test the technical knowledge and competence of the candidate; the oral is designed to evaluate intangible qualities, not readily measured otherwise, and to establish a list showing the relative fitness of each candidate – as measured against his competitors – for the position sought. Scoring is not on the basis of "right" and "wrong," but on a sliding scale of values ranging from "not passable" to "outstanding." As a matter of fact, it is possible to achieve a relatively low score without a single "incorrect" answer because of evident weakness in the qualities being measured.

Occasionally, an examination may consist entirely of an oral test – either an individual or a group oral. In such cases, information is sought concerning the technical knowledges and abilities of the candidate, since there has been no written examination for this purpose. More commonly, however, an oral test is used to supplement a written examination.

Who conducts interviews?

The composition of oral boards varies among different jurisdictions. In nearly all, a representative of the personnel department serves as chairman. One of the members of the board may be a representative of the department in which the candidate would work. In some cases, "outside experts" are used, and, frequently, a businessman or some other representative of the general public is asked to serve. Labor and management or other special groups may be represented. The aim is to secure the services of experts in the appropriate field.

However the board is composed, it is a good idea (and not at all improper or unethical) to ascertain in advance of the interview who the members are and what groups they represent. When you are introduced to them, you will have some idea of their backgrounds and interests, and at least you will not stutter and stammer over their names.

What should be done before the interview?

While knowledge about the board members is useful and takes some of the surprise element out of the interview, there is other preparation which is more substantive. It *is* possible to prepare for an oral interview – in several ways:

1) Keep a copy of your application and review it carefully before the interview

This may be the only document before the oral board, and the starting point of the interview. Know what education and experience you have listed there, and the sequence and dates of all of it. Sometimes the board will ask you to review the highlights of your experience for them; you should not have to hem and haw doing it.

2) Study the class specification and the examination announcement

Usually, the oral board has one or both of these to guide them. The qualities, characteristics or knowledges required by the position sought are stated in these documents. They offer valuable clues as to the nature of the oral interview. For example, if the job involves supervisory responsibilities, the announcement will usually indicate that knowledge of modern supervisory methods and the qualifications of the candidate as a supervisor will be tested. If so, you can expect such questions, frequently in the form of a hypothetical situation which you are expected to solve. NEVER go into an oral without knowledge of the duties and responsibilities of the job you seek.

3) Think through each qualification required

Try to visualize the kind of questions you would ask if you were a board member. How well could you answer them? Try especially to appraise your own knowledge and background in each area, *measured against the job sought*, and identify any areas in which you are weak. Be critical and realistic – do not flatter yourself.

4) Do some general reading in areas in which you feel you may be weak

For example, if the job involves supervision and your past experience has NOT, some general reading in supervisory methods and practices, particularly in the field of human relations, might be useful. Do NOT study agency procedures or detailed manuals. The oral board will be testing your understanding and capacity, not your memory.

5) Get a good night's sleep and watch your general health and mental attitude

You will want a clear head at the interview. Take care of a cold or any other minor ailment, and of course, no hangovers.

What should be done on the day of the interview?

Now comes the day of the interview itself. Give yourself plenty of time to get there. Plan to arrive somewhat ahead of the scheduled time, particularly if your appointment is in the fore part of the day. If a previous candidate fails to appear, the board might be ready for you a bit early. By early afternoon an oral board is almost invariably behind schedule if there are many candidates, and you may have to wait. Take along a book or magazine to read, or your application to review, but leave any extraneous material in the waiting room when you go in for your interview. In any event, relax and compose yourself.

The matter of dress is important. The board is forming impressions about you – from your experience, your manners, your attitude, and your appearance. Give your personal appearance careful attention. Dress your best, but not your flashiest. Choose conservative, appropriate clothing, and be sure it is immaculate. This is a business interview, and your appearance should indicate that you regard it as such. Besides, being well groomed and properly dressed will help boost your confidence.

Sooner or later, someone will call your name and escort you into the interview room. *This is it.* From here on you are on your own. It is too late for any more preparation. But remember, you asked for this opportunity to prove your fitness, and you are here because your request was granted.

What happens when you go in?

The usual sequence of events will be as follows: The clerk (who is often the board stenographer) will introduce you to the chairman of the oral board, who will introduce you to the other members of the board. Acknowledge the introductions before you sit down. Do not be surprised if you find a microphone facing you or a stenotypist sitting by. Oral interviews are usually recorded in the event of an appeal or other review.

Usually the chairman of the board will open the interview by reviewing the highlights of your education and work experience from your application – primarily for the benefit of the other members of the board, as well as to get the material into the record. Do not interrupt or comment unless there is an error or significant misinterpretation; if that is the case, do not hesitate. But do not quibble about insignificant matters. Also, he will usually ask you some question about your education, experience or your present job – partly to get you to start talking and to establish the interviewing "rapport." He may start the actual questioning, or turn it over to one of the other members. Frequently, each member undertakes the questioning on a particular area, one in which he is perhaps most competent, so you can expect each member to participate in the examination. Because time is limited, you may also expect some rather abrupt switches in the direction the questioning takes, so do not be upset by it. Normally, a board member will not pursue a single line of questioning unless he discovers a particular strength or weakness.

After each member has participated, the chairman will usually ask whether any member has any further questions, then will ask you if you have anything you wish to add. Unless you are expecting this question, it may floor you. Worse, it may start you off on an extended, extemporaneous speech. The board is not usually seeking more information. The question is principally to offer you a last opportunity to present further qualifications or to indicate that you have nothing to add. So, if you feel that a significant qualification or characteristic has been overlooked, it is proper to point it out in a sentence or so. Do not compliment the board on the thoroughness of their examination – they have been sketchy, and you know it. If you wish, merely say, "No thank you, I have nothing further to add." This is a point where you can "talk yourself out" of a good impression or fail to present an important bit of information. Remember, *you close the interview yourself.*

The chairman will then say, "That is all, Mr. ______, thank you." Do not be startled; the interview is over, and quicker than you think. Thank him, gather your belongings and take your leave. Save your sigh of relief for the other side of the door.

How to put your best foot forward

Throughout this entire process, you may feel that the board individually and collectively is trying to pierce your defenses, seek out your hidden weaknesses and embarrass and confuse you. Actually, this is not true. They are obliged to make an appraisal of your qualifications for the job you are seeking, and they want to see you in your best light. Remember, they must interview all candidates and a non-cooperative candidate may become a failure in spite of their best efforts to bring out his qualifications. Here are 15 suggestions that will help you:

1) Be natural – Keep your attitude confident, not cocky

If you are not confident that you can do the job, do not expect the board to be. Do not apologize for your weaknesses, try to bring out your strong points. The board is interested in a positive, not negative, presentation. Cockiness will antagonize any board member and make him wonder if you are covering up a weakness by a false show of strength.

2) Get comfortable, but don't lounge or sprawl

Sit erectly but not stiffly. A careless posture may lead the board to conclude that you are careless in other things, or at least that you are not impressed by the importance of the occasion. Either conclusion is natural, even if incorrect. Do not fuss with your clothing, a pencil or an ashtray. Your hands may occasionally be useful to emphasize a point; do not let them become a point of distraction.

3) Do not wisecrack or make small talk

This is a serious situation, and your attitude should show that you consider it as such. Further, the time of the board is limited – they do not want to waste it, and neither should you.

4) Do not exaggerate your experience or abilities

In the first place, from information in the application or other interviews and sources, the board may know more about you than you think. Secondly, you probably will not get away with it. An experienced board is rather adept at spotting such a situation, so do not take the chance.

5) If you know a board member, do not make a point of it, yet do not hide it

Certainly you are not fooling him, and probably not the other members of the board. Do not try to take advantage of your acquaintanceship – it will probably do you little good.

6) Do not dominate the interview

Let the board do that. They will give you the clues – do not assume that you have to do all the talking. Realize that the board has a number of questions to ask you, and do not try to take up all the interview time by showing off your extensive knowledge of the answer to the first one.

7) Be attentive

You only have 20 minutes or so, and you should keep your attention at its sharpest throughout. When a member is addressing a problem or question to you, give him your undivided attention. Address your reply principally to him, but do not exclude the other board members.

8) Do not interrupt

A board member may be stating a problem for you to analyze. He will ask you a question when the time comes. Let him state the problem, and wait for the question.

9) Make sure you understand the question

Do not try to answer until you are sure what the question is. If it is not clear, restate it in your own words or ask the board member to clarify it for you. However, do not haggle about minor elements.

10) Reply promptly but not hastily

A common entry on oral board rating sheets is "candidate responded readily," or "candidate hesitated in replies." Respond as promptly and quickly as you can, but do not jump to a hasty, ill-considered answer.

11) Do not be peremptory in your answers

A brief answer is proper – but do not fire your answer back. That is a losing game from your point of view. The board member can probably ask questions much faster than you can answer them.

12) Do not try to create the answer you think the board member wants

He is interested in what kind of mind you have and how it works – not in playing games. Furthermore, he can usually spot this practice and will actually grade you down on it.

13) Do not switch sides in your reply merely to agree with a board member

Frequently, a member will take a contrary position merely to draw you out and to see if you are willing and able to defend your point of view. Do not start a debate, yet do not surrender a good position. If a position is worth taking, it is worth defending.

14) Do not be afraid to admit an error in judgment if you are shown to be wrong

The board knows that you are forced to reply without any opportunity for careful consideration. Your answer may be demonstrably wrong. If so, admit it and get on with the interview.

15) Do not dwell at length on your present job

The opening question may relate to your present assignment. Answer the question but do not go into an extended discussion. You are being examined for a *new* job, not your present one. As a matter of fact, try to phrase ALL your answers in terms of the job for which you are being examined.

Basis of Rating

Probably you will forget most of these "do's" and "don'ts" when you walk into the oral interview room. Even remembering them all will not ensure you a passing grade. Perhaps you did not have the qualifications in the first place. But remembering them will help you to put your best foot forward, without treading on the toes of the board members.

Rumor and popular opinion to the contrary notwithstanding, an oral board wants you to make the best appearance possible. They know you are under pressure – but they also want to see how you respond to it as a guide to what your reaction would be under the pressures of the job you seek. They will be influenced by the degree of poise you display, the personal traits you show and the manner in which you respond.

ABOUT THIS BOOK

This book contains tests divided into Examination Sections. Go through each test, answering every question in the margin. We have also attached a sample answer sheet at the back of the book that can be removed and used. At the end of each test look at the answer key and check your answers. On the ones you got wrong, look at the right answer choice and learn. Do not fill in the answers first. Do not memorize the questions and answers, but understand the answer and principles involved. On your test, the questions will likely be different from the samples. Questions are changed and new ones added. If you understand these past questions you should have success with any changes that arise. Tests may consist of several types of questions. We have additional books on each subject should more study be advisable or necessary for you. Finally, the more you study, the better prepared you will be. This book is intended to be the last thing you study before you walk into the examination room. Prior study of relevant texts is also recommended. NLC publishes some of these in our Fundamental Series. Knowledge and good sense are important factors in passing your exam. Good luck also helps. So now study this Passbook, absorb the material contained within and take that knowledge into the examination. Then do your best to pass that exam.

EXAMINATION SECTION

EXAMINATION SECTION
TEST 1

DIRECTIONS: Each question or incomplete statement is followed by several suggested answers or completions. Select the one that BEST answers the question or completes the statement. *PRINT THE LETTER OF THE CORRECT ANSWER IN THE SPACE AT THE RIGHT.*

1. The relation between termites and their intestinal protozoa is an example of 1.____
   A. saprophytism B. parasitism
   C. symbiosis D. commensalism

2. The four basic tissues of vertebrates are 2.____
   A. connective, epithelial, blood, nerve
   B. epithelial, blood, nerve, muscle
   C. connective, epithelial, nerve, muscle
   D. connective, glandular, nerve, muscle

3. Of the following animals, regeneration of amputated parts can occur in all EXCEPT 3.____
   A. earthworms B. starfish
   C. snakes D. salamanders

4. The insect that feeds on the cottony-cushion scale is the 4.____
   A. ladybird beetle B. boll weevil
   C. Tachina fly D. Hessian fly

5. An organism with an *open circuit* system of circulation is the 5.____
   A. frog B. earthworm C. crayfish D. fish

6. Ringworm disease is caused by a 6.____
   A. lichen B. roundworm
   C. segmented worm D. fungus

7. The organism that causes typhus fever is a(n) 7.____
   A. fungus B. Rickettsia
   C. bacterium D. virus

8. Organizers are chemicals that influence 8.____
   A. differentiation B. mitosis
   C. fertilization D. maturation

9. The developing embryo of a mammal is protected by a liquid-filled sac called the 9.____
   A. placenta B. amnion C. uterus D. allantois

10. The haploid chromosome number in the fruit fly, Drosophila, is 10.____
    A. 4 B. 8 C. 24 D. 48

11. Typhus fever is a disease that is spread by an infected 11.____

A. tick
B. Anopheles mosquito
C. Aedes mosquito
D. body louse

12. The body plan of the spider is BEST described as 12.____

A. head, thorax, and abdomen
B. head and trunk
C. head-thorax and abdomen
D. head and thorax-abdomen

13. The frog differs from man in that it has no 13.____

A. salivary glands
B. adrenal glands
C. thyroid gland
D. pancreas

14. A frog's heart may be kept alive outside of the body by keeping it in 14.____

A. distilled water
B. Lugol's solution
C. Ringer's solution
D. sugar solution

15. The coelacanth is an important evolutionary link between the 15.____

A. coelenterates and the porifera
B. reptiles and the birds
C. amphibia and the reptiles
D. fish and the amphibia

16. Three of the following terms are applied to the larvae of various insects. The one that is NOT is 16.____

A. pupae
B. wrigglers
C. maggots
D. caterpillars

17. A worm that does not belong to the phylum Annelida is the 17.____

A. sandworm
B. leech
C. hookworm
D. earthworm

18. An example of a cold-blooded vertebrate is the 18.____

A. eel
B. whale
C. penguin
D. octopus

19. Three of the following are harmful insects. The one that is NOT is the 19.____

A. gypsy moth
B. codling moth
C. lady bug
D. cabbage butterfly

20. Three of the following are microscopic organisms. The one that is NOT is the 20.____

A. paramecium
B. euglena
C. ameba
D. sea urchin

21. Three of the following insects are capable of puncturing the human skin. The one that CANNOT is the 21.____

A. body louse
B. anopheles mosquito
C. flea
D. house fly

22. The structure which enabled the reptiles to break their link with the water is their 22.____

    A. skin B. egg
    C. bones D. sense organs

23. Of the following, the system likely to be MOST highly developed in internal parasites is the ______ system. 23.____

    A. digestive B. locomotor
    C. nervous D. reproductive

24. A striking difference between plants and animals is that plants 24.____

    A. excrete little or no CO_2
    B. do not respond to stimuli
    C. absorb little or no oxygen
    D. excrete little or no nitrogenous wastes

25. Mutations in fruit flies were FIRST produced with the use of x-rays by 25.____

    A. Muller B. Morgan C. Mendel D. Nichelson

KEY (CORRECT ANSWERS)

1.	C	11.	D
2.	C	12.	C
3.	C	13.	A
4.	A	14.	C
5.	C	15.	D
6.	D	16.	A
7.	B	17.	C
8.	A	18.	A
9.	B	19.	C
10.	A	20.	D

21.	D
22.	B
23.	D
24.	D
25.	A

TEST 2

DIRECTIONS: Each question or incomplete statement is followed by several suggested answers or completions. Select the one that BEST answers the question or completes the statement. *PRINT THE LETTER OF THE CORRECT ANSWER IN THE SPACE AT THE RIGHT.*

1. Earthworms in soil are useful because they 1.___

   A. destroy bacteria
   B. eat harmful substances
   C. increase the amount of carbon
   D. loosen the soil

2. Of the following, the one which does NOT spread disease by directly injecting the germ is 2.___

   A. anopheles mosquito
   B. body louse
   C. house fly
   D. tsetse fly

3. All of the following are insect poisons EXCEPT 3.___

   A. DDT B. Pyrethrum C. riboflavin D. rotenone

4. All of the following insects are harmful EXCEPT the 4.___

   A. codling moth
   B. Japanese beetle
   C. tent caterpillar
   D. lady bird beetle

5. Typhoid fever may be spread by all of the following EXCEPT 5.___

   A. milk
   B. water
   C. the housefly
   D. the mosquito

6. The bite of an insect is responsible for all of the following diseases EXCEPT 6.___

   A. malaria
   B. tetanus
   C. typhus
   D. yellow fever

7. The carriers of bubonic plague are known to be 7.___

   A. ticks
   B. tsetse flies
   C. lice
   D. houseflies
   E. fleas

Questions 8-10.

DIRECTIONS: Questions 8 through 10 refer to the drawing shown below.

8. The animal pictured is PROPERLY classified as a 8._____

A. black widow spider
B. carnivorous mammal
C. cold-blooded invertebrate
D. migratory waterfowl
E. clothes moth

9. This animal spent an earlier period of its existence as a 9._____

A. bacterium B. virus C. drone
D. maggot E. fossil

10. It is probable that this animal is an active carrier of the germs that cause 10._____

A. malaria B. tuberculosis C. diabetes
D. rickets E. yellow fever

11. A maggot is the larva of a(n) 11._____

A. mosquito B. butterfly C. fly
D. ant E. bee

12. It is believed by scientists that worker bees indicate to their hive mates the direction and distance of a supply of food by 12._____

A. strokes of their antennae
B. buzzing sounds
C. an oriented dance in front of the hive
D. none of the above

13. Three of the plants listed below are insectivorous. The one that is NOT is the 13._____

A. Venus's flytrap B. Pitcher plant
C. Indian pipe D. Sundew

14. That molluscs and annelids arose from a common ancestral type is indicated by the 14._____

A. bipinnaria larva B. specific fossil evidence
C. tornaria larva D. trocophore larva

15. A developmental stage associated with incomplete metamorphosis in insects is called 15._____

A. larva B. pupa C. nymph D. maggot

16. An example of a hermaphroditic animal is the 16._____

A. centipede B. earthworm C. spider D. stickleback

17. Two insects that belong to different orders are 17._____

A. cricket and grasshopper B. bee and ant
C. cicada and dragonfly D. butterfly and moth

18. Of the following chromosome types, the one that would produce a female Drosophila fly is 18._____

A. XT B. XX C. XO D. OY

19. Some plant lice are protected by certain ants that derive a sweet juice from the plant lice. This type of biological association is known as 19.____

A. commensalism
B. parasitism
C. saprozoic nutrition
D. symbiosis

20. All malarial parasites belong to the genus 20.____

A. Clostridium
B. Plasmodium
C. Rhizobium
D. Spirillum

21. A vector of a number of disease-producing trypanosomes is the 21.____

A. body louse
B. flea
C. leech
D. tsetse fly

22. The usual portal of entry of the hookworm is 22.____

A. exposed skin
B. bite of an insect
C. nasal passages
D. ingestion of improperly cooked food

23. The respiratory system of insects includes a complex network of air tubes called 23.____

A. alveoli
B. bronchial tubes
C. spiracles
D. tracheae

24. The organism that causes the disease trichinosis is a member of the phylum that includes 24.____

A. annelids
B. flatworms
C. roundworms
D. fungi

25. An insect that destroys a large number of plant lice is the 25.____

A. hawkmoth
B. ichneumon fly
C. ladybird beetle (ladybug)
D. robber fly

KEY (CORRECT ANSWERS)

1. D
2. C
3. C
4. D
5. D
6. B
7. E
8. C
9. D
10. B
11. C
12. C
13. C
14. D
15. C
16. B
17. C
18. B
19. D
20. B
21. D
22. A
23. D
24. C
25. C

TEST 3

DIRECTIONS: Each question or incomplete statement is followed by several suggested answers or completions. Select the one that BEST answers the question or completes the statement. *PRINT THE LETTER OF THE CORRECT ANSWER IN THE SPACE AT THE RIGHT.*

1. Mosaic vision is characteristic of the 1.___

   A. chameleon
   B. dragon fly
   C. leopard frog
   D. nighthawk

2. An example of close interdependence in a biotic environment is exhibited by 2.___

   A. oak and moth
   B. corn and cornborer
   C. maple and bee
   D. termite and protozoa

3. A Drosophila fly having two X chromosomes and three sets of autosomes is called 3.___

   A. female
   B. intersex
   C. supermale
   D. superfemale

4. A dye generally used to demonstrate the giant chromosomes of Drosophila larvae is 4.___

   A. aceto carmine
   B. brom thymol blue
   C. eosin
   D. indophenol

5. A disease caused by a trypanosome that is transmitted by a dipteran insect is 5.___

   A. filariasis
   B. malaria
   C. schistosomiasis
   D. sleeping sickness

6. Parasites that may grow in humans and block the flow of fluids in lymph ducts are the 6.___

   A. ascarids
   B. filaria
   C. malaria
   D. trichinella

7. To destroy sucking insects on laboratory plants, it is BEST to use 7.___

   A. flit
   B. rotenone
   C. soap solution
   D. 2-4D

8. To separate female drosophila for crossing, choose flies which have a _______ abdomen. 8.___

   A. black-tipped
   B. broad-spotted
   C. broad-striped
   D. small-spotted

9. An example of symbiotic relationship is the 9.___

   A. dodder
   B. earthworm
   C. lichen
   D. tapeworm

10. A disease which may be prevented by wearing shoes is caused by 10.___

    A. hookworm
    B. mealy worm
    C. tapeworm
    D. trichina worm

11. The chromosome number of the fruitfly is normally 8. It might be 16 if it were not kept down by the process of 11.___

    A. conjugation
    B. maturation
    C. mitosis
    D. parthenogenesis

12. Muller is known for his work in producing mutations MAINLY in 12.____

A. corn
B. fruitflies
C. Neurospora
D. planaria

13. The brontosaurus was MOST closely related to our present 13.____

A. frog
B. salamander
C. snake
D. porpoise

14. Of the following animals, the one possessing marked variable temperature is the 14.____

A. snake B. elephant C. horse D. canary

15. Air passes into and out of the insect body by way of 15.____

A. skin B. lungs C. gills D. tracheae

16. The female drosophila can be distinguished from the male by noting that 16.____

A. the female possesses a fringe of black bristles in the uppermost joint of the first pair of legs
B. the tip of the abdomen in the female is elongated and more rounded
C. the first pair of legs in the male is no longer than in the female
D. the female has dark bristles in the third pair of legs

17. Lichens are composites, consisting of fungi and 17.____

A. bacteria B. algae C. mosses D. protozoa

18. Stock cultures drosophila should be kept in rooms which have no greater range in temperature than from 18.____

A. 15-20°C B. 20-25°C C. 25-30°C D. 35-40°C

19. The relationship between termites and ciliates is known as 19.____

A. mutualism
B. parasitism
C. commensalism
D. holophytism

20. The relation of termites and their intestinal flagellates is a classic example of 20.____

A. amensalism
B. commensalism
C. parasitism
D. symbiosis

21. Fossils are USUALLY found in rocks classified as 21.____

A. metamorphic
B. igneous
C. conglomerate
D. sedimentary

22. Before it reaches the adult stage, the grasshopper must FIRST be a 22.____

A. larva B. pupa C. nymph D. grub

23. A close relative of the earthworm is the 23.____

A. ascaris B. planaria C. leech D. tapeworm

24. A self-fertilizing hermaphroditic organism is the 24.____

A. tapeworm B. honeybee C. snail D. oyster

25. Chromosomes much larger than normal may be seen in preparations of 25.____

A. stems of tall pea plants
B. salivary glands of drosophila larvae
C. pituitary glands
D. roots of onions

KEY (CORRECT ANSWERS)

1.	B	11.	B
2.	D	12.	B
3.	B	13.	C
4.	A	14.	A
5.	D	15.	D
6.	B	16.	B
7.	B	17.	B
8.	C	18.	A
9.	C	19.	B
10.	A	20.	D

21. D
22. C
23. C
24. A
25. B

TEST 4

DIRECTIONS: Each question or incomplete statement is followed by several suggested answers or completions. Select the one that BEST answers the question or completes the statement. *PRINT THE LETTER OF THE CORRECT ANSWER IN THE SPACE AT THE RIGHT.*

1. An example of an insect predator is the 1.____

   A. termite
   B. bumblebee
   C. grasshopper
   D. doddlebug

2. The effect of x-rays on the rate of mutation in drosophila was discovered by 2.____

   A. Muller
   B. Morgan
   C. Weismann
   D. deVries

3. The insect with the LONGEST life cycle is the 3.____

   A. praying mantis
   B. cicada
   C. honeybee
   D. termite

4. The sex of honeybees is controlled by 4.____

   A. the drone
   B. the queen
   C. chance
   D. the temperature

5. Flatworms are USUALLY found in ponds or streams 5.____

   A. on the underside of objects
   B. free-swimming
   C. on the upper side of objects
   D. on the north side of objects

6. In certain insects, the sex formula is XX-XO. In both males and females, there are 22 autosomes. The chromosome number found in sperm cells is 6.____

   A. 11
   B. 12
   C. 11 and 12
   D. 23

7. In drosophila, the sex of a fly of the genotype XY is determined PRIMARILY by genes located on the 7.____

   A. autosomes and X-chromosomes
   B. Y-chromosome
   C. autosomes
   D. autosomes and X-chromosome

8. Mites and ticks are MOST closely related to 8.____

   A. true bugs
   B. click beetles
   C. silverfish
   D. spiders

9. Nightcrawlers are 9.____

   A. crayfish
   B. eel larvae
   C. earthworms
   D. moth larvae

10. Respiratory organs found in insects are called 10.____

A. gills
B. book lungs
C. lungs
D. tracheae

11. Of the following, the way in which arachnids resemble other arthropods is that they have 11.____

A. four pairs of walking legs
B. no compound eyes
C. no antennae
D. jointed exoskeletons

12. An open-type blood circulation is found in all of the following EXCEPT the 12.____

A. earthworms
B. insects
C. Crustacea
D. mollusks

13. The excretory organs of insects are called 13.____

A. green glands
B. nephridia
C. Malpighian tubules
D. glomeruli

14. The osmotic effect of water in the paramecium is controlled by the 14.____

A. cell membrane
B. pellicle
C. macronucleus
D. contractile vacuoles

15. Which one of the following is used as a mold inhibitor in the preparation of a Drosophila culture medium? 15.____

A. Ninhydrin
B. Nipagen
C. Pyridoxin
D. Glycine

16. Which one of the following is arranged in a CORRECT sequence? 16.____

A. Family, genus, phylum, order
B. Order, family, species, variety
C. Phylum, order, genus, family
D. Kingdom, order, species, genus

17. Which one of the following pairs of animal groups consists of two groups belonging to the same phylum? 17.____

A. Myriapods and leeches
B. Slugs and earthworms
C. Sand dollars and clams
D. Tarantulas and barnacles

18. In vertebrates, blood cells which carry oxygen are known as 18.____

A. leucocytes
B. erythrocytes
C. thrombocytes
D. monocytes

19. Gynandromorphs appear in Drosophila because of 19.____

A. cytoplasmic inheritance
B. autosomes
C. loss of X-chromosomes
D. gain of Y-chromosome

20. In which one of the following were giant chromosomes discovered? 20.____

A. Fruit fly
B. Man
C. Garden pea
D. Four o'clock

21. Recently, a *screw-worm* pest has been eliminated from the southeastern part of the United States by means of 21.____

A. DDT
B. 2-4D
C. natural enemies
D. irradiation of males

22. The corpora allata are small glandular organs found in which one of the following? 22.____

A. Moths
B. Crayfish
C. Mammals
D. Salamanders

23. Reproduction by parthenogenesis is a naturally recurring process in the reproduction of which one of the following? 23.____

A. Bees
B. Earthworms
C. Moths
D. Japanese beetles

24. An invertebrate that goes through metamorphosis in water is the 24.____

A. eel
B. mosquito
C. sea horse
D. salamander

25. A natural enemy imported to destroy San Jose scale on orange trees was a type of 25.____

A. digger wasp
B. dragon fly
C. ladybird beetle
D. praying mantis

KEY (CORRECT ANSWERS)

1. D
2. A
3. B
4. B
5. A
6. C
7. D
8. D
9. C
10. D
11. D
12. A
13. C
14. D
15. B
16. B
17. D
18. B
19. C
20. A
21. D
22. A
23. A
24. B
25. C

EXAMINATION SECTION
TEST 1

DIRECTIONS: In the following questions, there is a key term in capital letters followed by four alternatives. From among the four alternatives, choose the one which is MOST CLOSELY associated with the key term. *PRINT THE LETTER OF THE CORRECT ANSWER IN THE SPACE AT THE RIGHT.*

1. SOFT TICKS 1.____
   A. Order Hemiptera
   B. Order Diptera
   C. Class Arachnida
   D. Order Chilopoda

2. BUGS (TRUE BUGS) 2.____
   A. Order Hemiptera
   B. Order Anoplura
   C. Order Siphonaptera
   D. Order Chilopoda

3. SANDFLIES 3.____
   A. Order Anoplura
   B. Order Diptera
   C. Order Hemiptera
   D. Class Arachnida

4. HARD TICKS 4.____
   A. Family Argasidae
   B. Family Ixodidae
   C. Class Arachnida
   D. Order Diptera

5. MOSQUITOES 5.____
   A. Order Chilopoda
   B. Order Diptera
   C. Order Hemiptera
   D. Order Siphonaptera

6. CENTIPEDES 6.____
   A. Order Hemiptera
   B. Order Diptera
   C. Order Anoplura
   D. Order Chilopoda

7. MIDGES 7.____
   A. Order Chilopoda
   B. Order Diptera
   C. Order Anoplura
   D. Order Siphonaptera

8. LICE 8.____
   A. Order Hemiptera
   B. Order Chilopoda
   C. Order Anoplura
   D. Family Ixodidae

9. FLIES 9.____
   A. Order Siphonaptera
   B. Order Diptera
   C. Class Arachnida
   D. Family Argasidae

10. CYCLOPS 10.____
    A. Scabies
    B. Scrub typhus
    C. Texas cattle fever
    D. Guinea worm

KEY (CORRECT ANSWERS)

1. C
2. A
3. B
4. C
5. B
6. D
7. C
8. A
9. B
10. A

TEST 2

DIRECTIONS: In the following questions, there is a key term in capital letters followed by four alternatives. From among the four alternatives, choose the one which is MOST CLOSELY associated with the key term. *PRINT THE LETTER OF THE CORRECT ANSWER IN THE SPACE AT THE RIGHT.*

1. XENOPSYLLA CHEOPSIS 1.____
   A. Rat flea
   B. Cat flea
   C. Human flea
   D. Human bot fly

2. GLOSSINA SP. 2.____
   A. Myasis-producing fly
   B. Human bot fly
   C. Tsetse-fly
   D. Cat flea

3. CORDYLOBIA ANTHROPOPHAGA 3.____
   A. Tsetse-fly
   B. Myasis-producing fly
   C. Human bot fly
   D. Human flea

4. PULEX IRRITANS 4.____
   A. Human flea
   B. Cat flea
   C. Rat flea
   D. Tsetse-fly

5. SARCOPHAGA HAEMORRHOIDALIS 5.____
   A. Rat flea
   B. Myasis-producing fly
   C. Tsetse-fly
   D. Cat flea

6. CTENOCEPHALIDES FELIS 6.____
   A. Rat flea
   B. Cat flea
   C. Human flea
   D. Human bot fly

7. DERMATOBIA HOMINIS 7.____
   A. Rat flea
   B. Myasis-producing fly
   C. Tsetse-fly
   D. Human bot fly

8. FLEAS 8.____
   A. Order Siphonaptera
   B. Order Chilopoda
   C. Order Hemiptera
   D. Order Diptera

9. SPIDERS 9.____
   A. Family Ixodidae
   B. Family Argasidae
   C. Class Arachnida
   D. Order Anoplura

10. SCORPIONS 10.____
    A. Order Hemiptera
    B. Order Diptera
    C. Class Arachnida
    D. Family Ixodidae

KEY (CORRECT ANSWERS)

1. A
2. C
3. B
4. A
5. B
6. B
7. D
8. A
9. C
10. C

TEST 3

DIRECTIONS: In the following questions, there is a key term in capital letters followed by four alternatives. From among the four alternatives, choose the one which is MOST CLOSELY associated with the key term. *PRINT THE LETTER OF THE CORRECT ANSWER IN THE SPACE AT THE RIGHT.*

1. ORNITHORODORUS MOUBATA 1.____
   A. Guinea worm
   B. Texas cattle fever
   C. Relapsing fever
   D. Scabies

2. TROMBICULA SP. 2.____
   A. Scabies
   B. Guinea worm
   C. Scrub typhus
   D. Large fish tapeworm

3. LACTRODECTUS MACTANS 3.____
   A. Spider whose venom contains a powerful toxin
   B. Rocky Mountain spotted fever
   C. Relapsing fever
   D. Large fish tapeworm

4. SARCOPTES SCABIEI 4.____
   A. Rocky Mountain spotted fever
   B. Scabies
   C. Scrub typhus
   D. Guinea worm

5. BOOPHILUS ANNULATUS 5.____
   A. Large fish tapeworm
   B. Scabies
   C. Relapsing fever
   D. Texas cattle fever

6. DIAPTOMUS 6.____
   A. Large fish tapeworm
   B. Guinea worm
   C. Scrub typhus
   D. Scabies

7. DERMACENTOR ANDERSONI 7.____
   A. Rocky Mountain spotted fever
   B. Texas cattle fever
   C. Relapsing fever
   D. Scabies

8. RHODNIUS PROLIXUS 8.____
   A. Bed bug
   B. Kissing bug
   C. Crab louse
   D. Head and body louse

9. CULEX SP. 9.____
   A. Crab louse
   B. Bed bug
   C. Mosquito
   D. Kissing bug

10. CIMEX LECTULARIUS 10.____

A. Bed bug
B. Kissing bug
C. Mosquito
D. Head and body louse

KEY (CORRECT ANSWERS)

1.	C	6.	A
2.	C	7.	A
3.	A	8.	B
4.	B	9.	C
5.	D	10.	A

TEST 4

DIRECTIONS: In the following questions, there is a key term in capital letters followed by four alternatives. From among the four alternatives, choose the one which is MOST CLOSELY associated with the key term. *PRINT THE LETTER OF THE CORRECT ANSWER IN THE SPACE AT THE RIGHT.*

1. PEDICULUS HUMANUS 1.____
   A. Bed bug
   B. Crab louse
   C. Kissing bug
   D. Head and body louse

2. ANOPHELES SP. 2.____
   A. Crab louse
   B. Head and body louse
   C. Mosquito
   D. Bed bug

3. TRIATOMA SP. 3.____
   A. Crab louse
   B. Kissing bug
   C. Mosquito
   D. Bed bug

4. PHTHIRIUS PUBIS 4.____
   A. Crab louse
   B. Bed bug
   C. Kissing bug
   D. Mosquito

5. AEDES SP. 5.____
   A. Head and body louse
   B. Bed bug
   C. Mosquito
   D. Crab louse

6. SIMULIUM DAMNOSUM 6.____
   A. Black widow spider
   B. Sand fly
   C. Black fly
   D. Horse fly

7. TABANUS SP. 7.____
   A. Stable fly
   B. Black widow spider
   C. Black fly
   D. Horse fly

8. BUTHUS SP. 8.____
   A. Scorpion
   B. Gnat
   C. Stable fly
   D. Horse fly

9. LATRODECTUS MACTANS 9.____
   A. Black widow spider
   B. Gnat
   C. Scorpion
   D. Stable fly

10. CULICOIDES SP. 10.____
    A. Scorpion
    B. Gnat
    C. Sand fly
    D. Horse fly

KEY (CORRECT ANSWERS)

1.	D	6.	C
2.	C	7.	D
3.	B	8.	A
4.	A	9.	A
5.	C	10.	B

TEST 5

DIRECTIONS: In the following questions, there is a key term in capital letters followed by four alternatives. From among the four alternatives, choose the one which is MOST CLOSELY associated with the key term. *PRINT THE LETTER OF THE CORRECT ANSWER IN THE SPACE AT THE RIGHT.*

1. CHRYSOPS SP. 1.____
   A. Scorpion
   B. Gnat
   C. Stable fly
   D. Horse fly

2. PHLEBOTOMUS SP. 2.____
   A. Black fly
   B. Sand fly
   C. Black widow spider
   D. Stable fly

3. STOMOXYS CALCITRANS 3.____
   A. Stable fly
   B. Sand fly
   C. Black fly
   D. Horse fly

4. CENTRUROIDES SP. 4.____
   A. Scorpion
   B. Black widow spider
   C. Sand fly
   D. Gnat

5. ARGAS SP. 5.____
   A. Soft tick
   B. Hard tick
   C. Chigger (red bug)
   D. Texas cattle tick

6. AMBLYOMA AMERICANUM 6.____
   A. Common wood tick
   B. Hard tick
   C. *Lone star* tick
   D. Itch mite

7. DEMODEX FOLLICULORUM 7.____
   A. Texas cattle tick
   B. Chigger (red bug)
   C. Follicle mite
   D. Itch mite

8. ORNITHODORUS SP. 8.____
   A. Soft tick
   B. Follicle mite
   C. *Lone star* tick
   D. Common wood tick

9. BOOPHILUS ANULATUS 9.____
   A. Common wood tick
   B. Hard tick
   C. Soft tick
   D. Itch mite

10. TROMBICULA SP. 10.____
   A. Chigger (red bug)
   B. Follicle mite
   C. Itch mite
   D. Common wood tick

KEY (CORRECT ANSWERS)

1. D
2. B
3. A
4. A
5. A
6. B
7. C
8. A
9. B
10. A

TEST 6

DIRECTIONS: In the following questions, there is a key term in capital letters followed by four alternatives. From among the four alternatives, choose the one which is MOST CLOSELY associated with the key term. *PRINT THE LETTER OF THE CORRECT ANSWER IN THE SPACE AT THE RIGHT.*

1. DERMACENTOR ANDERSONI 1.____
   A. Soft tick
   B. Hard tick
   C. Common wood tick
   D. *Lone star* tick

2. SARCOPTES SCABIEI 2.____
   A. Follicle mite
   B. Itch mite
   C. Common wood tick
   D. Texas cattle tick

3. ANOPHELES MOSQUITO 3.____
   A. Bancroft's filariasis
   B. Malaria
   C. Trench fever
   D. Chagas' disease

4. CULEX MOSQUITO 4.____
   A. Bancroft's filiariasis
   B. Epidemic typhus
   C. Bubonic plague
   D. Dipylidiasis

5. AEDES MOSQUITO 5.____
   A. Bancroft's filariasis
   B. Trench fever
   C. Chagas' disease
   D. Epidemic typhus

6. PEDICULUS HUMANUS (LOUSE) 6.____
   A. Trench fever
   B. Epidemic typhus
   C. Diplidiasis
   D. Bubonic plague

7. DOG FLEA 7.____
   A. Chagas' disease
   B. Dipylidiasis
   C. Malaria
   D. Bancroft's filariasis

8. RAT FLEA 8.____
   A. Malaria
   B. Trench fever
   C. Bubonic plague
   D. Epidemic typhus

9. KISSING BUG (ASSASSIN BUG) 9.____
   A. Trench fever
   B. Epidemic typhus
   C. Dipylidiasis
   D. Chagas' disease

10. AEDES MOSQUITO 10.____
    A. Yellow fever
    B. Dengue fever
    C. Leishmaniasis
    D. Loaiasis

KEY (CORRECT ANSWERS)

1. B
2. C
3. B
4. A
5. A
6. B
7. B
8. C
9. D
10. A

TEST 7

DIRECTIONS: In the following questions, there is a key term in capital letters followed by four alternatives. From among the four alternatives, choose the one which is MOST CLOSELY associated with the key term. *PRINT THE LETTER OF THE CORRECT ANSWER IN THE SPACE AT THE RIGHT.*

1. CHRYSOPS SP. 1.____

   A. African trypanosomiasis
   B. Tularemia and anthrax
   C. Yellow fever
   D. Loaiasis

2. HOUSE FLY 2.____

   A. Onchocerciasis
   B. Amebic and bacillary dysenteries
   C. Acanthocheilonema and Mansonella infections
   D. Cutaneous myasis in man

3. PHLEBOTOMUS SAND FLY 3.____

   A. Leishmaniasis
   B. Yellow fever
   C. Tularemia and anthrax
   D. Onchocerciasis

4. TSETSE-FLY 4.____

   A. African trypanosomiasis
   B. Dengue fever
   C. Cutaneous myasis in man
   D. Leishmaniasis

5. BLACK FLY (SIMULIUM SP.) 5.____

   A. Acanthocheilonema and Mansonella infections
   B. Amebic and bacillary dysenteries
   C. Onchocerciasis
   D. African trypanosomiasis

6. HORSE BOT FLY 6.____

   A. Loaiasis
   B. Cutaneous myasis in man
   C. Tularemia and anthrax
   D. Yellow fever

7. GNAT (CULICOIDES SP.) 7.____

   A. Onchocerciasis
   B. Loaiasis
   C. Yellow fever
   D. Acanthocheilonema and Mansonella infections

8. TABANUS SP. 8.____

   A. Tularemia and anthrax
   B. African trypanosomiasis
   C. Cutaneous myasis in man
   D. Dengue fever

9. BANCROFT'S FILARIASIS 9.____

   A. Anopheles mosquito
   B. Rat flea
   C. Culex mosquito
   D. Dog flea

10. HUMAN FLEA 10.____

A. Xenopsylla cheopis
B. Glossina sp.
C. Dermatobia hominis
D. Pulex irritans

KEY (CORRECT ANSWERS)

1. D
2. B
3. A
4. A
5. C
6. B
7. D
8. A
9. C
10. D

EXAMINATION SECTION
TEST 1

DIRECTIONS: Each question or incomplete statement is followed by several suggested answers or completions. Select the one that BEST answers the question or completes the statement. *PRINT THE LETTER OF THE CORRECT ANSWER IN THE SPACE AT THE RIGHT.*

1. In his dealings with tenants, an exterminator will find that tenants will usually cooperate willingly to carry out a legitimate request if they 1.____

   A. are asked to do it at the time the exterminator leaves
   B. have lived in the building for more than a year
   C. fear the exterminator's official authority
   D. understand the reason for the request

2. When planning his work schedule at a housing project, an exterminator should FIRST determine 2.____

   A. how the work should be done
   B. what work has to be done
   C. when the work must be completed
   D. what equipment and supplies are needed

3. Assume that your foreman and you disagree as to how a certain bait should be applied. You think your way is better.
   Under these circumstances, you should generally 3.____

   A. be frank and tell your foreman why you think your way is better
   B. discuss the matter with a higher ranking member of the pest control section
   C. ignore what the foreman says and do it your way
   D. postpone further discussion until your foreman is in a better mood

4. Your foreman reprimands you for creating a safety hazard by not closing containers of pesticides when they are not in use. However, you know that you closed the containers after you had finished using them. These containers were left open by another exterminator.
   It would be MOST advisable for you to close the containers and take which one of the following actions? 4.____

   A. Find out who the exterminator was and get this person to tell the foreman what really happened
   B. Complain to other exterminators about your foreman's lack of fairness and judgment
   C. Notify your foreman's boss of the unjust accusation
   D. Tell the foreman that the containers were left open by another exterminator

5. Of the following, the MOST advisable thing for you to do if your foreman asks you to operate a piece of equipment you do not know how to handle is to 5.____

   A. ask one of the other exterminators if he could do the job for you this time
   B. ask one of the men in the maintenance department of the housing development to show you how to use it

C. tell your foreman that you do not know how to operate it
D. try it out until you figure out how it works

6. Suppose that you receive a shipment of 50 rat snap-traps from a hardware dealer. When you unpack them, you find that in seven of the traps, the springs and triggers are not fastened securely to the trap because there are deep cracks in the wood where they are attached. 6.____
Of the following, it would be MOST advisable for you to

A. cover the cracks with masking tape and put all the traps on the stock shelf in the shop
B. list only 43 traps in your stock records and put the damaged traps aside for salvage
C. send all the traps back to the hardware dealer
D. tell your foreman about the damaged traps

7. Of the following, the MAIN purpose of a safety training program is to 7.____

A. fix the blame for accidents
B. describe accidents which have occurred
C. hold the Housing Exterminators responsible for unsafe working conditions
D. make the Housing Exterminators aware of the basic causes of accidents

8. When administering first aid to a person suffering from shock as a result of an accident, of the following, it is MOST important to 8.____

A. cover the person and keep him warm
B. apply artificial respiration
C. prop him up in a sitting position
D. massage the person in order to aid blood circulation

9. Assume you have just been appointed. You notice that certain equipment which is assigned to you is defective and that use of this equipment may eventually result in unnecessary costs and perhaps injury to you. 9.____
The BEST thing for you to do is to

A. speak to the maintenance men in the project about repairing the equipment
B. discuss the matter with your foreman
C. mind your own business since you have just been appointed
D. speak to other exterminators and find out if they had any experience with defective equipment

10. Assume you are working in a project building and one of the housing caretakers has just been seriously injured in an accident in the slop sink room. 10.____
Your FIRST concern should be to

A. help the injured man
B. find the cause of the accident
C. report the accident to your foreman
D. report the accident to the caretaker's boss

11. Assume a mass of extension cords plugged into one outlet in a pest control shop results in overloading the electrical circuit and causes a fire. 11.____
Which of the following types of extinguisher should be used to put out the fire?

A. Carbon dioxide (CO_2)
B. Water
C. Soda acid
D. Carbon tetrachloride

12. Manufacturers of chemical pesticides usually recommend that special precautions be taken when the chemicals are used. 12.____
Of the following, which one would a manufacturer be LEAST likely to recommend?

A. Wear leather gloves
B. Wear a respirator
C. Wear safety goggles
D. Have a first aid kit available

Questions 13-15.

DIRECTIONS: Questions 13 through 15 consist of groups of statements that have to do with safety precautions and procedures. Choose the statement in each group that is NOT correct.

13. 13.____
A. The label on the original container of the pesticide should be read before each use.
B. Pest control equipment should be cleaned regularly.
C. Whenever there is a choice of chemicals, the chemical which is less hazardous to humans should be used at all times.
D. For the transfer of concentrates from drums, either threaded taps or drum pumps should be used.

14. 14.____
A. Do not use a petroleum base on an asphalt tile floor.
B. Do not spray oil base sprays on material colored with oil soluble dyes.
C. Do not use respirators.
D. Do not use pesticides which are highly poisonous to mammals.

15. The following statements deal with disposal of empty gallon metal containers which held highly toxic organic phosphate insecticides. 15.____

A. Do not reuse these containers.
B. Pour one pint of water into the empty container, add bicarbonate of soda, and bury the container of rinse solution at least 18 inches below ground.
C. Wet all inner surfaces with the proper rinse solution.
D. Punch holes in the top and bottom of the can, crush the can and bury deeply in an isolated location.

KEY (CORRECT ANSWERS)

1. D
2. B
3. A
4. D
5. C
6. D
7. D
8. A
9. B
10. A
11. A
12. A
13. C
14. C
15. B

TEST 2

DIRECTIONS: Each question or incomplete statement is followed by several suggested answers or completions. Select the one that BEST answers the question or completes the statement. *PRINT THE LETTER OF THE CORRECT ANSWER IN THE SPACE AT THE RIGHT.*

1. There are 22 boxes of rat mix in a certain pest control shop. If each box contains 7 1/2 pounds of rat mix, the TOTAL amount of rat mix in the shop is ______ pounds. 1.____

   A. 165 B. 172 1/2 C. 180 D. 182 1/2

2. A pest control shop has a supply of 26 one-gallon cans of insecticide. If the exterminator works 5 days a week and uses 32 ounces of the liquid a day, the number of work-weeks this supply of insecticide will last is MOST NEARLY 2.____

   A. 10 B. 20 C. 28 D. 32

3. A certain supplier packs two dozen mouse traps to a box. If the exterminator gets a delivery of 20 boxes and finds that two of these boxes are half-full, the TOTAL number of traps the exterminator received from this supplier is 3.____

   A. 408 B. 432 C. 456 D. 480

4. Assume that a truck which contains a shipment of pesticides ia parked outside your exterminating shop. You are able to unload the truck in one hour. How long would it take four exterminators, starting at the same time and working at the same rate as you, to unload four trucks similar to the one you unloaded? 4.____

   A. 15 minutes B. 1 hour
   C. 2 hours D. 4 hours

5. A certain building in a housing development has 142 apartments. It takes on exterminator an average of six minutes to treat one apartment. At that rate, approximately how long should it take him to treat ALL 142 apartments? ______ hours. 5.____

   A. 2 B. 14 C. 24 D. 85

6. A crate contains 3 pieces of pesticide equipment weighing 73, 84, and 47 pounds, respectively. If the crate is lifted by 4 exterminators, each lifting one corner of the crate, the average number of pounds, in addition to the weight of the crate, lifted by each of the exterminators is 6.____

   A. 51 B. 65 C. 71 D. 78

7. A stack of cartons containing pesticides is 10 cartons long, 9 cartons wide, and 5 cartons high. The number of cartons in the stack is 7.____

   A. 24 B. 55 C. 95 D. 450

8. Assume that you have bags of corn meal, each of the same weight. The total weight of 25 bags is 125 pounds. 8.___
How many of these bags would it take to make a TOTAL weight of 50 pounds?

A. 2 B. 5 C. 6 D. 10

9. You are working in the sub-basement of a project building and the foreman tells you to get two boards from the maintenance shop to stand on. One of the boards is 5 yards long and the other is 3 1/2 feet long. 9.___
The TOTAL length, in feet, of the two boards is

A. 8 1/2 B. 9 1/2 C. 17 1/2 D. 18 1/2

10. Three hundred plastic bags of rat mix, each bag weighing four ounces, are packed in a carton. The carton weighed one pound before the rat mix was packed in it. 10.___
The TOTAL weight of the filled carton is ______ pounds.

A. 37 1/2 B. 38 1/2 C. 75 D. 76

11. In a certain project, an exterminator is told to spray 120 apartments, 40 apartments, 20 basements, 80 apartments, 50 storerooms, and 10 compactors. 11.___
The TOTAL number of apartments which should be sprayed is

A. 240 B. 260 C. 310 D. 320

Questions 12-15.

DIRECTIONS: Questions 12 through 15 are to be answered SOLELY on the basis of the following information.

The insects you, as a housing exterminators will control are just a minute fraction of the millions which inhabit the world. Man does well to hold his own in the face of the constant pressures that insects continue to exert upon him. Not only are the total numbers tremendous, but the number of individual kinds, or species, certainly exceeds 800,000—a number greater than that of all other animals combined. Many of these are beneficial but some are especially competitive with man. Not only are insects numerous, but they are among the most adaptable of all animals. In their many forms, they are fitted for almost any specific way of life. Their adaptability, combined with their tremendous rate of reproduction, gives insects an unequaled potential for survival!

The food of insects includes almost anything that can be eaten by any other animal as well as many things which cannot even be digested by any other animals. Most insects do not harm the products of man or carry diseases harmful to him; however, many do carry diseases and others feed on his food and manufactured goods. Some are adapted to living only in open areas while others are able to live in extremely confined spaces. All of these factors combined, make the insects a group of animals having many members which are a nuisance to man and thus of great importance to the housing exterminator.

The control of insects requires an understanding of their way of life. Thus, it is necessary for the housing exterminator to understand the anatomy of the insect, its method of growth, the time it takes for the insect to grow from egg to adult, its habits, the stage of its life history in which it causes damage, its food, and its common living places. In order to obtain the best control, it is especially important to be able to identify correctly the specific insect involved because without this knowledge, it is impossible to prescribe a proper treatment.

12. Which one of the following is a CORRECT statement about the insect population of the world, according to the above paragraph? 12.____
The
    A. total number of insects is less than the total number of all other animals combined
    B. number of species of insects is greater than the number of species of all other animals combined
    C. total number of harmful insects is greater than the total number of those which are not harmful
    D. number of species of harmless insects is less than the number of species of those which are harmful

13. Insects will be controlled MOST efficiently if the housing exterminator 13.____
    A. understands why the insects are so numerous
    B. knows what insects he is dealing with
    C. sees if the insects compete with man
    D. is able to identify the food which the insects digest

14. According to the above passage, insects are of importance to an exterminator PRIMARILY because they 14.____
    A. can be annoying, destructive, and harmful to man
    B. are able to thrive in very small spaces
    C. cause damage during their growth stages
    D. are so adaptable that they can adjust to any environment

15. According to the above passage, insects can eat 15.____
    A. everything that any other living thing can eat
    B. man's food and things which he makes
    C. anything which other animals can't digest
    D. only food and food products

KEY (CORRECT ANSWERS)

1. A	6. A
2. B	7. D
3. C	8. D
4. B	9. D
5. B	10. D

11. A
12. B
13. B
14. A
15. B

EXAMINATION SECTION
TEST 1

DIRECTIONS: Each question or incomplete statement is followed by several suggested answers or completions. Select the one that BEST answers the question or completes the statement. *PRINT THE LETTER OF THE CORRECT ANSWER IN THE SPACE AT THE RIGHT.*

Questions 1-4.

DIRECTIONS: Questions 1 through 4 are to be answered on the basis of the information provided in the paragraph below.

Rodent control must be planned carefully in order to insure its success. This means that more knowledge is needed about the habits and favorite breeding places of Domestic Rats, than any other kind. A favorite breeding place for Domestic Rats is known to be in old or badly constructed buildings. Rats find these buildings very comfortable for making nests. However, the only way to gain this kind of detailed knowledge about rats is through careful study.

1. According to the above paragraph, rats find comfortable nesting places 1.____

   A. in old buildings
   B. in pipes
   C. on roofs
   D. in sewers

2. The paragraph states that the BEST way to learn all about the favorite nesting places of rats is by 2.____

   A. asking people
   B. careful study
   C. using traps
   D. watching ratholes

3. According to the paragraph, in order to insure the success of rodent control, it is necessary to 3.____

   A. design better bait
   B. give out more information
   C. plan carefully
   D. use pesticides

4. The paragraph states that the MOST important rats to study are ______ rats. 4.____

   A. African
   B. Asian
   C. Domestic
   D. European

Questions 5-8.

DIRECTIONS: Questions 5 through 8 are to be answered on the basis of the following paragraph.

A few people who live in old tenements have the bad habit of throwing garbage out of their windows, especially if there is an empty lot near their building. Sometimes the garbage is food, sometimes the garbage is half-empty soda cans. Sometimes the garbage is a little bit of both mixed together. These people just don't care about keeping the lot clean.

5. The paragraph states that throwing garbage out of windows is a 5.____

A. bad habit
B. dangerous thing to do
C. good thing to do
D. good way to feed rats

6. According to the paragraph, an empty lot next to an old tenement is sometimes used as a place to 6.____

A. hold local gang meetings
B. play ball
C. throw garbage
D. walk dogs

7. According to the paragraph, which of the following throw garbage out of their windows? 7.____

A. Nobody
B. Everybody
C. Most people
D. Some people

8. According to the paragraph, the kinds of garbage thrown out of windows are 8.____

A. candy and cigarette butts
B. food and half-empty soda cans
C. fruit and vegetables
D. rice and bread

Questions 9-12.

DIRECTIONS: Questions 9 through 12 are to be answered on the basis of the following paragraph.

The game that is recognised all over the world as an all-American game is the game of baseball. As a matter of fact, baseball heroes like Joe DiMaggio, Willie Mays, and Babe Ruth, were as famous in their day as movie stars Robert Redford, Paul Newman, and Clint Eastwood are now. All these men have had the experience of being mobbed by fans whenever they put in an appearance anywhere in the world. Such unusual popularity makes it possible for stars like these to earn at least as much money off the job as on the job. It didn't take manufacturers and advertising men long to discover that their sales of shaving lotion, for instance, increased when they got famous stars to advertise their product for them on radio and television.

9. According to the paragraph, baseball is known everywhere as a(n) _______ game. 9.____

A. all-American
B. fast
C. unusual
D. tough

10. According to the paragraph, being so well known means that it is possible for people like Willie Mays and Babe Ruth to 10.____

A. ask for anything and get it
B. make as much money off the job as on it
C. travel anywhere free of charge
D. watch any game free of charge

11. According to the paragraph, which of the following are known all over the world? 11.____

A. Baseball heroes
B. Advertising men
C. Manufacturers
D. Basketball heroes

12. According to the paragraph, it is possible to sell much more shaving lotion on television and radio if 12.____

    A. the commercials are in color instead of black and white
    B. you can get a prize with each bottle of shaving lotion
    C. the shaving lotion makes you smell nicer than usual
    D. the shaving lotion is advertised by famous stars

Questions 13-16.

DIRECTIONS: Questions 13 through 16 are to be answered on the basis of the following paragraph.

People are very suspicious of all strangers who knock at their door. For this reason, every pest control aide, whether man or woman, must carry an identification card at all times on the job. These cards are issued by the agency the aide works for. The aide's picture is on the card. The aide 's name is typed in, and the aide 's signature is written on the line below. The name, address, and telephone number of the agency issuing the card is also printed on it. Once the aide shows this ID card to prove his or her identity, the tenant's time should not be taken up with small talk. The tenant should be told briefly what pest control means. The aide should be polite and ready to answer any questions the tenant may have on the subject. Then, the aide should thank the tenant for listening and say goodbye.

13. According to the above paragraph, when she visits tenants, the one item a pest control aide must ALWAYS carry with her is a(n) 13.____

    A. badge
    B. driver's license
    C. identification card
    D. watch

14. According to the paragraph, a pest control aide is supposed to talk to each tenant he visits 14.____

    A. at length about the agency
    B. briefly about pest control
    C. at length about family matters
    D. briefly about social security

15. According to the paragraph, the item that does NOT appear on an ID card is the 15.____

    A. address of the agency
    B. name of the agency
    C. signature of the aide
    D. social security number of the aide

16. According to the paragraph, a pest control aide carries an identification card because he must 16.____

    A. prove to tenants who he is
    B. provide the tenants with the agency's address
    C. provide the tenant with the agency's telephone number
    D. save the tenant's time

Questions 17-20.

DIRECTIONS: Questions 17 through 20 are to be answered on the basis of the following paragraph.

Very early on a summer's morning, the nicest thing to look at is a beach, before the swimmers arrive. Usually all the litter has been picked up from the sand by the Park Department clean-up crew. Everything is quiet. All you can hear are the waves breaking, and the sea gulls calling to each other. The beach opens to the public at 10 A.M. Long before that time, however, long lines of eager men, women, and children have driven up to the entrance. They form long lines that wind around the beach waiting for the signal to move.

17. According to the paragraph, before 10 A.M., long lines are formed that are made up of 17.____

A. cars
B. clean-up crews
C. men, women, and children
D. Park Department trucks

18. The season referred to in the above paragraph is 18.____

A. fall B. summer C. winter D. spring

19. The place the paragraph is describing is a 19.____

A. beach
B. park
C. golf course
D. tennis court

20. According to the paragraph, one of the things you notice early in the morning is that 20.____

A. radios are playing
B. swimmers are there
C. the sand is dirty
D. the litter is gone

Questions 21-30.

DIRECTIONS: In Questions 21 through 30, select the answer which means MOST NEARLY the SAME as the capitalized word in the sentence.

21. He received a large REWARD. 21.____
In this sentence, the word REWARD means

A. capture
B. recompense
C. key
D. praise

22. The aide was asked to TRANSMIT a message. In this sentence, the word TRANSMIT means 22.____

A. change B. send C. take D. type

23. The pest control aide REQUESTED the tenant to call the Health Department. 23.____
In this sentence, the word REQUESTED means the pest control aide

A. asked B. helped C. informed D. warned

24. The driver had to RETURN the Health Department's truck. In this sentence, the word RETURN means 24.____

A. borrow B. fix C. give back D. load up

25. The aide discussed the PURPOSE of the visit. In this sentence, the word PURPOSE means 25.____

A. date B. hour C. need D. reason,

26. The tenant SUSPECTED the aide who knocked at her door. In this sentence, the word SUSPECTED means 26.____

A. answered
B. called
C. distrusted
D. welcomed

27. The aide was POSITIVE that the child hit her. In this sentence, the word POSITIVE means 27.____

A. annoyed B. certain C. sorry D. surprised

28. The tenant DECLINED to call the Health Department. In this sentence, the word DECLINED means 28.____

A. agreed B. decided C. refused D. wanted

29. The aide ARRIVED on time. 29.____
In this sentence, the word ARRIVED means

A. awoke B. came C. left D. delayed

30. The salesman had to DELIVER books to each person he visited. 30.____
In this sentence, the word DELIVER means

A. give B. lend C. mail D. sell

KEY (CORRECT ANSWERS)

1.	A	11.	A	21.	B
2.	B	12.	D	22.	B
3.	C	13.	C	23.	A
4.	C	14.	B	24.	C
5.	A	15.	D	25.	D
6.	C	16.	A	26.	C
7.	D	17.	C	27.	B
8.	B	18.	B	28.	C
9.	A	19.	A	29.	B
10.	B	20.	D	30.	A

TEST 2

DIRECTIONS: Each question or incomplete statement is followed by several suggested answers or completions. Select the one that BEST answers the question or completes the statement. *PRINT THE LETTER OF THE CORRECT ANSWER IN THE SPACE AT THE RIGHT.*

Questions 1-10.

DIRECTIONS: In Questions 1 through 10, pick the word that means MOST NEARLY the OPPOSITE of the capitalize word in the sentence.

1. It is possible to CONSTRUCT a rat-proof home. The opposite of CONSTRUCT is 1.____

   A. build B. erect C. plant D. wreck

2. The pest control aide had to REPAIR the flat tire. The opposite of the word REPAIR is 2.____

   A. destroy B. fix C. mend D. patch

3. The pest control aide tried to SHOUT the answer. The opposite of the word SHOUT is 3.____

   A. scream B. shriek C. whisper D. yell

4. Daily VISITS are the best. 4.____
   The opposite of the word VISITS is

   A. absences B. exercises C. lessons D. trials

5. It is important to ARRIVE early in the morning. The opposite of the word ARRIVE is 5.____

   A. climb B. descend C. enter D. leave

6. Jorge is a group LEADER. 6.____
   The opposite of the word LEADER is

   A. boss B. chief C. follower D. overseer

7. The EXTERIOR of the house needs painting. 7.____
   The opposite of the word EXTERIOR is

   A. inside B. outdoors C. outside D. surface

8. He CONCEDED the victory. 8.____
   The opposite of the word CONCEDED is

   A. admitted B. denied C. granted D. reported

9. He watched the team BEGIN. 9.____
   The opposite of the word BEGIN is

   A. end B. fail C. gather D. win

10. Your handwriting is ILLEGIBLE. 10.____
    The opposite of the word ILLEGIBLE is

    A. clear B. confused C. jumbled D. unclear

Questions 11-15.

DIRECTIONS: Questions 11 through 15 are to be answered by following the instructions given in each question. Note that 5 possible answers have been given for these questions ONLY. Therefore, for these questions, your choice may be A, B, C, D, or E. .

11. Add: 11.____

$$12\frac{1}{2}$$
$$2\frac{1}{4}$$
$$3\frac{1}{4}$$

The CORRECT answer is

A. 17 B. 174 C. 174 D. 17 3/4 E. 18

12. Subtract: 12.____

150
-80

The CORRECT answer is

A. 70 B. 80 C. 130 D. 150 E. 230

13. After cleaning up some lots in the East Bronx, five cleanup crews loaded the following amounts of garbage on trucks: 13.____
Crew No. 1 loaded 2 1/4 tons
Crew No. 2 loaded 3 tons
Crew No. 3 loaded 1 1/4 tons
Crew No. 4 loaded 2 1/4 tons
Crew No. 5 loaded 1/2 ton
The TOTAL number of tons of garbage loaded was

A. 8 B. 8 1/4 C. 8 3/4 D. 9 E. 9 1/4

14. Subtract: 14.____

17 3/4
- 7 1/4

The CORRECT answer is

A. 7 1/2 B. 10 1/2 C. 14 1/4 D. 17 3/4 E. 25

15. Yesterday, Tom and Bill each received 10 leaflets about rat control. Each supermarket in the neighborhood was supposed to receive one of these leaflets. When the day was over, Tom had 8 leaflets left. Bill had no leaflets left. How many supermarkets got leaflets yesterday? 15.____

A. 8 B. 10 C. 12 D. 18 E. 20

Questions 16-20.

DIRECTIONS: Questions 16 through 20 are to be answered ONLY on the basis of the information in the following statement and chart, DAILY WORK REPORT FORM (Chart A).

Assume that you are a member of the Pest Control Truck Crew Number 1. Julio Rivera is your Crew Chief. The crew is supposed to report to work at nine o'clock in the morning, Since you are the first to show up, *at ten minutes before nine, on 5/24 Rivera asks you to help him out by filling in the Daily Work Report Form for him. Driver Hal Williams shows up at nine, and Driver Rick Smith shows up ten minutes after Williams.*

DAILY WORK REPORT FORM (Chart A)

<table>
<tr><td>Block #1

Crew No.</td><td colspan="2">Block #2

Date</td></tr>
<tr><td>Block #3 TRUCKS IN USE
Truck # ____
____
____
____
____
____
____
____
____
____</td><td>Block #4 DRIVER'S NAME
____
____
____
____
____
____
____
____
____</td><td>Block #5 <u>TIME OF ARRIVAL</u>
<u>A.M.</u> <u>P.M.</u></td></tr>
<tr><td>Block #6 <u>TRUCKS OUT OF ORDER</u>
____
____
____
____
#</td><td>Block #7 <u>ADDRESS OF CLEAN-UP SITE</u>
No.____

Street____</td><td>Block #8

Borough

Block #9

Signature of Crew Chief</td></tr>
</table>

16. According to the above statement, the entry that belongs in Block #9 is 16.____

    A. Julio Rivera
    B. June Stevens
    C. Jim Watson
    D. Hal Williams

17. According to the above statement, the entry that should be made in Block #2 is 17.____

    A. 9:00 A.M.
    B. 9:10 P.M.
    C. 5/24
    D. 7/24

18. The names of Hal Williams and Rick Smith should appear in Block # 18.____

    A. 4
    B. 6
    C. 7
    D. 9

19. Rick Smith's time of arrival should be entered in Block #5 as ______ A.M. 19.____

A. 8:50 B. 8:55 C. 9:00 D. 9:10

20. According to the statement, the entry that should be made in Block #1 is 20.____

A. zero B. one C. 5/24 D. 6/24

Questions 21-23.

DIRECTIONS: Questions 21 through 23 are to be answered on the basis of the statement shown below. Use DAILY WORK REPORT FORM (Chart A) on Page 3 as a guide.

Pete Marberg showed up at a quarter after nine, in the morning, but his truck, No. 22632441, was in the garage for repairs. Steve Marino showed up a half hour after Pete. He was assigned truck No. 6342003, which was in working order.

21. According to the above statement, truck No. 22632441 should be entered in Block # 21.____

A. 3 B. 4 C. 6 D. 8

22. According to the above statement, Steve Marino showed up at 22.____

A. 9:00 A.M. B. 9:15 A.M. C. 9:30 P.M. D. 9:45 A.M.

23. According to the above statement, Steve Marino's truck number belongs in Block #3. The number entered there should be 23.____

A. 22632441 B. 6342003 C. 6432003 D. 26232441

Questions 24-30.

DIRECTIONS: Questions 24 through 30 are to be answered ONLY on basis of the information in the statements above ea question and the following chart, DAILY GARBAGE COLLECTION REPORT (Chart B).

DAILY GARBAGE COLLECTION REPORT C Chart B)

Block #1	Block #2	Block #3	Block #4	Block #5
No. of Trucks Used For Collection	Address of Garbage Pick-Up	Amount of Garbage Collected	Amount of Garbage Unloaded	Hours During Which Garbage Was Unloaded
#456	45 Southwest	1/2 ton	1/2 ton	From 7 AM To 8 AM
		Block #6 otal Amount of Garbage Collected By All Trucks	Block #7 Total Amount of Garbage Unloaded By All Trucks	Block #8 Total Amount of Time Spent Unloading Of All Trucks
TOTALS ______				

24. Truck # 2437752 started unloading garbage at ten o'clock Monday morning and finished unloading its garbage that afternoon. The clock looked like this when the job was done. 24.____

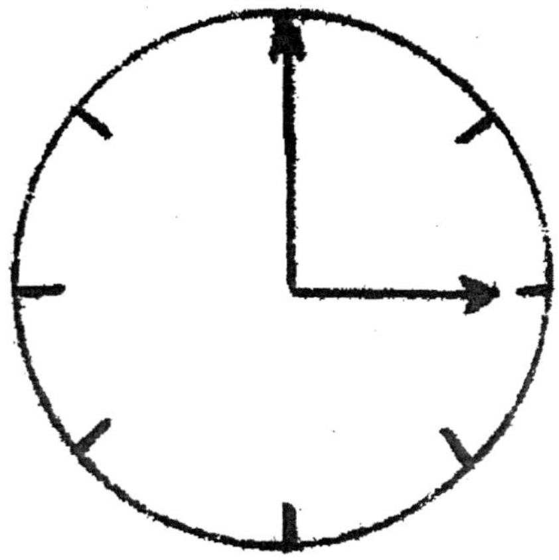

The time entries that should be recorded in Block #5 are

A. 10 A.M. and 12:15 P.M.
B. 10 P.M. and 12:30 A.M.
C. 10 P.M. and 12:00 A.M.
D. 10 A.M. and 3:00 P.M.

25. Truck # 8967432 had to pick up a load of garbage from 911 South Avenue. It took the crew until 11:00 A.M. to load the garbage. 25.____
According to this statement, the item 911 South Avenue should be entered in Block #

A. 1 B. 2 C. 3 D. 4

26. On Tuesday, truck # 124356 unloaded 4 ton of garbage, truck # 2437752 unloaded J ton of garbage, and truck # 435126 unloaded 1/2 ton of garbage. 26.____
The TOTAL amount of garbage unloaded by the three trucks on Tuesday should be entered in Block #

A. 3 B. 4 C. 5 D. 8

27. On Wednesday, it took truck # 4050607 from 2 P.M. to 6 P.M. to unload 1 ton of garbage. It took truck # 7040650 from 1 P.M. to 2 P.M. to unload 1/4 ton of garbage. These were the only trucks working that day. 27.____
The TOTAL amount of time it took for both trucks to unload garbage was ______ hours.

A. 5 B. 6 C. 7 D. 8

28. The amount of garbage collected by one truck should be entered in the DAILY GARBAGE COLLECTION REPORT FORM in Block # 28.____

A. 3 B. 6 C. 7 D. 8

29. Truck # 557799010 reported to 1020 Hudson River Alley to pick up garbage from an empty lot. 29.____
This information should be entered in the DAILY GARBAGE COLLECTION REPORT FORM in Block # ______ and Block # ______ .

A. 1; 4 B. 2; 5 C. 1; 2 D. 2; 3

30. It took the Pest Control Truck crew from 8 in the morning to 12 noon to unload the garbage it collected the night before. 30.____
This information should be entered in the DAILY GARBAGE COLLECTION REPORT FORM under Block #

A. 4 B. 5 C. 6 D. 7

KEY (CORRECT ANSWERS)

1.	D	11.	E	21.	C
2.	A	12.	A	22.	D
3.	C	13.	E	23.	B
4.	A	14.	B	24.	D
5.	D	15.	C	25.	B
6.	C	16.	A	26.	B
7.	A	17.	C	27.	A
8.	B	18.	A	28.	A
9.	A	19.	D	29.	C
10.	A	20.	B	30.	B

EXAMINATION SECTION
TEST 1

DIRECTIONS: Each question or incomplete statement is followed by several suggested answers or completions. Select the one that BEST answers the question or completes the statement. *PRINT THE LETTER OF THE CORRECT ANSWER IN THE SPACE AT THE RIGHT.*

Questions 1-20.

DIRECTIONS: Questions 1 through 20 are to be answered SOLELY on the basis of the information contained in the following passage, the CREW CHIEF'S REPORT, and the HEALTH INSPECTOR'S REPORT.

Block 421 is located in a heavily industrialized part of the city. There are eight six-story apartment buildings on the block. Each of these apartment buildings contains 24 five-room apartments.

When a child living in one of the apartment buildings on block 421 needs medical care, he is taken to the privately-owned child clinic a few blocks away. Dr. Stone, the director of the clinic, recently learned that 25 of the 40 children treated by his clinic for rat bites within the past year lived on block 421 at the time they were bitten. He was so concerned about the conditions on that block that he wrote a letter to the Department of Health requesting that a pest control team be sent there to help reduce the rodent threat to children living on block 421.

In response to Dr. Stone"s letter, a team of 20 pest control workers under the supervision of a crew chief named John Angelo was sent to work on block 421. They remained there for a period of four weeks. During the first week, the pest control workers cleared the garbage from the streets and backyards of the block while the crew chief inspected all of the apartment buildings there to deter-mine the extent of the rodent infestation. The crew chief and his staff spent the second week cleaning up the basements, halls, and stairs of each apartment building on the block and filling in whatever rat holes they found there. During the third week, they placed rat poison in the basements and apartments of each building. They spent the fourth week instructing the tenants of block 421 on how to keep their apartments clean of rats. Just before they left the block for the last time, Mr. Angelo placed a sign in the lobby of each apartment house which gave the name, address, and telephone number of the director of the nearest rodent control office. All of the tenants were urged to phone this number if the problem with rodents developed again.

After three months had passed and nobody from block 421 called the rodent control office to complain of a rat problem, the health department concluded that rodents were no longer a problem on block 421 and they concentrated their efforts elsewhere. Then they received a letter from Dr. Stone which said that children living on block 421 were coming to the clinic with rat bites again. He demanded that the health department conduct a follow-up inspection to determine just how successful they had been in removing rats from the block.

In response to Dr. Stone's second letter, a health inspector named Harry Rosen was sent to block 421 to make a follow-up inspection. He discovered that the block showed as many signs of rodent infestation as it had when the pest control team had first inspected the block, prior to cleaning it up. Mr. Rosen recommended that a second pest control team be sent to block 421 and that a more intensive community education program be conducted after the block was once again cleaned up and the rodents were once again removed.

Reports were prepared by both the crew chief and the health inspector concerning their inspection of block 421. Copies of each of these reports labeled the CREW CHIEF'S REPORT and the HEALTH INSPECTOR'S REPORT follow.

CREW CHIEF'S REPORT

DATE INSPECTION COMPLETED: August 17 PREPARED BY: John Angelo, Crew Chief

BLOCK INSPECTED: 421

Bldg. Address	No. of Apts. Needing Exterminating;	Conditions Found in Public Areas	Conditions Found in Apartments
5 Pier St.	12	Halls and stairs badly littered. Rat burrows in backvard.	Kitchen floors food stained. Rat holes in baseboards.
11 Pier St.	7	Basement entrance blocked by debris. Rat droppings in basement .	Uncovered garbage in kitchens. Garbage frequently thrown out windows.
6 Water Rd.	20	Holes in basement floors need filling. Rat holes behind boiler.	Rat droppings in foyers. Uncovered garbage in kitchens .
10 Water Rd.	14	Rat droppings in basement. Dead rats spotted in basement .	Kitchen floors food stained. Rat holes in baseboards.
5 Bay Rd.	4	Fire escapes littered. Backyard filled with sewage due to backun.	Garbage frequently thrown out windows. Uncovered garbage in kitchens.
9 Bay Rd.	24	Basement entrance blocked by debris. Rat holes behind boiler.	Uncovered food containers on counters. Kitchen floors food stained.
10 Canal St.	10	Rat droppings in basement. Halls and stairs badly littered .	Rat holes in baseboards. Uncovered garbage in Kitchen
14 Canal St.	14	Rat tracks seen in bacKyard. Basement entrance blocked bv debris.	Rat holes in baseboards. Uncovered food containers on counters.

HEALTH INSPECTOR'S REPORT

DATE INSPECTION COMPLETED: April 17 PREPARED BY: Harry Rosen, Health Inspector

BLOCK INSPECTED: 421

Bldg. Address	No. of Apts. Needing Exterminating	Conditions Found in Public Areas	Conditions Found in Apartments
5 Pier St.	8	Rat droppings in basement. Basement entrance blocked bv debris.	Rat holes in baseboards. Uncovered garbage in kitchens.
11 Pier St.	12	Rat holes behind boiler. Holes in basement floor need filling.	Rat holes in baseboards. Kitchen floors food stained.
6 Water Rd.	24	Rat holes behind boiler. Halls and stairs badly littered.	Holes in kitchen walls behind sink. Rat holes in baseboards.
10 Water Rd.	14	Basement entrance blocked by debris. Traces of old rat bait found in cellar.	Garbage frequently thrown out windows. Uncovered garbage in kitchens.
5 Bay Rd.	6	Rat droppings in basement. Dead rats found in basement.	Holes in kitchen walls behind sink. Rat droppings in fovers.
9 Bay Rd.	17	Basement entrance blocked by debris. Fire escapes littered.	Rat holes in baseboards. Uncovered food containers on counters.
10 Canal St.	14	Entrance to building badly littered. Rat holes behind boiler.	Kitchen floors food stained. Uncovered garbage in kitchens.
14 Canal St.	21	Rat droppings in basement. Halls and stairs badly littered.	Garbage frequently thrown; out windows. Rat holes in baseboards.

1. The total number of apartments on block 421 is 1.____

   A. 144 B. 192 C. 960 D. 1152

2. When children living on block 421 need health care, they are taken to the 2.____

   A. child clinic
   B. family doctor
   C. Health Department
   D. hospital

3. Pest control workers were sent to block 421 because 3.____

   A. the block is in a highly industrialized part of the city
   B. numerous tenants complained of rodent infestation on the block
   C. Dr. Stone asked that they be sent there
   D. forty children living on that block had been bitten by rats

4. The pest control workers left block 421 after they had been working there for _______ week(s). 4.___

A. one B. two C. three D. four

5. An inspection of the buildings on block 421 was INITIALLY made by 5.___

A. Dr. Stone
B. a pest control worker
C. Harry Rosen
D. John Angelo

6. In treating the rodent-infested buildings on block 421, the pest control workers placed rat poison 6.___

A. in basements and apartments
B. in building lobbies
C. in halls and on stairs
D. on streets and in backyards

7. Just before they left block 421, the pest control workers 7.___

A. cleared the buildings and surrounding areas of garbage
B. placed rat bait wherever it was needed
C. placed a sign in the lobby of each building
D. told the tenants how they could keep their apartments clean

8. After the pest control team finished its job, the Health Department thought that block 421 was clean of rodents because 8.___

A. they had received no further complaints from Dr. Stone
B. they had received no complaints from the tenants who lived there
C. no more children on the block were being bitten by rodents
D. a follow-up inspection had found no further evidence of rodents there

9. When a health inspector from the Department of Health conducted a follow-up inspection of block 421, he learned that 9.___

A. the block was still rat infested
B. the block was now free of rodents
C. all of the buildings now had their own exterminators
D. some of the buildings had been abandoned

10. The health inspector who conducted the follow-up inspection recommended that 10.___

A. the building owners hire their own exterminators
B. a second pest control team be sent to work on block 421
C. block 421 needed no further pest control work
D. all of the buildings on block 421 be demolished

11. The health inspector found the largest number of apartments needing exterminating in the building at 11.___

A. 6 Water Road
B. 10 Water Road
C. 9 Bay Road
D. 14 Canal Street

12. In which of the following buildings did the crew chief find the fewest number of apartments needing exterminating? 12.___

A. 5 Pier Street
B. 11 Pier Street
C. 9 Bay Road
D. 10 Canal Street

13. In which of the following buildings did Harry Rosen find food stained kitchen floors? 13.____

A. 5 Pier Street
B. 6 Water Road
C. 9 Bay Road
D. 10 Canal Street

14. In which of the following buildings did John Angelo find the basement entrance blocked by debris? 14.____

A. 5 Pier Street
B. 10 Water Road
C. 9 Bay Road
D. 10 Canal Street

15. In which of the following buildings did the health inspector find littered fire escapes? 15.____

A. 11 Pier Street
B. 6 Water Road
C. 5 Bay Road
D. 9 Bay Road

16. Harry Rosen found rat holes in the baseboards of apartments in all of the following buildings EXCEPT 16.____

A. 5 Pier Street
B. 10 Water Road
C. 9 Bay Road
D. 14 Canal Street

17. How much time passed between the time the crew chief completed his inspection and the time the health inspector completed his? 17.____

A. 4 weeks B. 4 months C. 6 months D. 8 months

18. The only building in which both Harry Rosen and John Angelo found 14 apartments needing extermination was 18.____

A. 11 Pier Street
B. 10 Water Road
C. 10 Canal Street
D. 14 Canal Street

19. The total number of apartments which the crew chief found in need of exterminating was 19.____

A. 76 B. 87 C. 105 D. 116

20. The building in which Harry Rosen found 12 apartments needing exterminating was 20.____

A. 5 Pier Street
B. 11 Pier Street
C. 6 Water Road
D. 14 Canal Street

KEY (CORRECT ANSWERS)

1. B	11. A
2. A	12. B
3. C	13. D
4. D	14. C
5. D	15. D
6. A	16. B
7. C	17. D
8. B	18. B
9. A	19. C
10. B	20. B

TEST 2

DIRECTIONS: Each question or incomplete statement is followed by several suggested answers or completions. Select the one that BEST answers the question or completes the statement. *PRINT THE LETTER OF THE CORRECT ANSWER IN THE SPACE AT THE RIGHT.*

Questions 1-10.

DIRECTIONS: Questions 1 through 10 are to be answered SOLELY on the basis of the information contained in the following passage and refer to entries that would be made on the FIELD VISIT REPORT form that follows the passage.

On March 6, a crew composed of five Community Service Aides and three Pest Control Aides, under the supervision of a Crew Chief (Pest Control), made a field visit to inspect several residential buildings and a vacant lot. The purpose of the visit was to check for exposed refuse and signs of rats, mice, and insects. If conditions needed correction, they were to recommend the actions that should be taken.

The crew was driven in a department car to the first inspection site, an apartment house at 124 Grand Street, arriving at 11:30 A.M. When the crew members inspected the apartment house, they discovered rats and holes in the baseboards in several of the apartments. The landlord had not placed enough bait boxes in the basement. The Crew Chief recommended that an exterminator be scheduled to treat the building. The crew left the building at 12:05 P.M. and walked to the next inspection site at 129 Grand Street.

The crew arrived at the second site at 12:10 P.M. and left at 12:40 P.M. Because the crew found rats and roaches in the building, the Crew Chief immediately called the office and made arrangements for an exterminator to treat the building that afternoon. The Crew Chief recommended that the building should be re-inspected the following week to see if the exterminating had been successful.

The crew workers walked to the next inspection site, a vacant lot on Lucke Street, across the street from an apartment building at 350 Lucke Street. They observed that refuse covered much of the area of the vacant lot. The Crew Chief recommended that a clean-up team be scheduled to remove refuse from the lot.

The crew's last inspection of the day was a building at 300 Lucke Street. They walked to this site, arrived at 1:00 P.M., and stayed for an hour. They inspected several apartments in the building to see if a recent extermination had been successful. Upon seeing that no further work was needed at the site, they returned to their office by subway.

The Crew Chief arrived at the office at 3:00 P.M. and made out the following FIELD VISIT REPORT form:

FIELD VISIT REPORT FORM

1. Date ____________________
2. Time Arrived at First Site ____________________
3. Purpose of Field Visit ____________________
____
____
4. Number of Persons in Crew (Not including Crew Chief (Pest Control) __________
____
5. Transportation ____________________
6. Number of Sites Visited ____________________
7. Addresses of Sites Visited ____________________
____
____
____
8. Conditions Noted ____________________
____
____
____
9. Recommendations ____________________
____
____
____
10. Arrangements Made by Crew Chief While in the Field __________
____
____
____
11. Time Left Last Site ____________________

1. Which of the following should be entered on line 2? 1.____

A. 11:30 A.M.
B. 12:05 P.M.
C. 12:10 P.M.
D. 12:40 P.M.

2. Which of the following should be entered on line 3? 2.____

A. Exterminate apartment buildings that have rats and mice
B. Examine various sites for exposed refuse and signs of rats, mice, and insects
C. Inspect work done by clean-up team
D. Clean up lots that are covered with refuse

3. The number that should be entered on line 4 is 3.____

A. 3 B. 5 C. 8 D. 9

4. Which of the following should be entered on line 6? 4.____

A. 3 B. 4 C. 5 D. 6

5. Each of the following should be entered on line 7 EXCEPT 5._____

A. 124 Grand Street
B. 129 Grand Street
C. 300 Lucke Street
D. 350 Lucke Street

6. Each of the following should be entered on line 8 EXCEPT the presence of 6._____

A. holes in the baseboards at 124 Grand Street
B. insects, rats, and mice at 300 Lucke Street
C. refuse at the vacant lot on Lucke Street
D. rats and roaches at 129 Grand Street

7. Which of the following should be entered on line 5? 7._____

A. Department car to first site, subway between sites
B. Subway to first site, walked between sites
C. Walked to first site, department car between sites
D. Department car to first site, walked between sites

8. All of the following should be entered on line 9 EXCEPT: 8._____

A. Extermination at 124 Grand Street to remove rats
B. Clean-up at the lot on Lucke Street to remove refuse
C. Follow-up visit at 129 Grand Street to determine success of extermination
D. Clean-up building at 300 Lucke Street to end infestation

9. Which of the following should be entered on line 10? 9._____

A. Extermination of building at 129 Grand Street
B. Extermination of building at 124 Grand Street
C. Clean-up of lot on Lucke Street
D. Clean-up of building at 300 Lucke Street

10. Which of the following should be entered on line 11? 10._____
_______ P.M.

A. 12:40
B. 1:00
C. 2:00
D. 3:00

KEY (CORRECT ANSWERS)

1. A	6. B
2. B	7. D
3. C	8. D
4. B	9. A
5. D	10. C

EXAMINATION SECTION
TEST 1

DIRECTIONS: Each question or incomplete statement is followed by several suggested answers or completions. Select the one that BEST answers the question or completes the statement. *PRINT THE LETTER OF THE CORRECT ANSWER IN THE SPACE AT THE RIGHT.*

1. You are following up on the inspections which have been made by one of your inspectors whose work is usually satisfactory. You visit an establishment recently inspected by him and note several violations of the Code which the inspector had failed to report. You discuss the matter with the inspector who becomes highly indignant and insists that the establishment complied with the provisions of the Code at the time of his inspection. Under the circumstances, it would be MOST advisable for you to state that 1.____

   A. a report of the incident, submitted to the borough chief, will be included in the inspector's personnel file
   B. at the time of your visit the premises did not comply with some of the provisions of the Code
   C. future failure to report violations of the Code will be regarded as presumptive evidence of collusion
   D. you will recommend a transfer or termination of the employment of the inspector if the situation occurs again

2. You have been assigned to spotcheck the daily report of an inspector. The inspector has indicated that he inspected a certain establishment at 3 P.M. The owner of the establishment insists that the inspector inspected his premises at 11 A.M. Of the following courses of action, you should FIRST 2.____

   A. interview the owners of the establishment visited by the inspector before and after the establishment in question
   B. secure the inspector's daily report for the previous day and check every stop on that report
   C. telephone the inspector to determine the actual time of his visit
   D. write a formal memorandum to the borough chief regarding the incident

3. You are investigating the complaint made by the owner of an establishment who alleged that an inspector spoke to him in a loud and disrespectful manner while inspecting his premises. You interview the complainant and ask him if he has any witnesses to support his complaint. He tells you that he does not. You note that the inspector found several violations of the Code in the course of his inspection of the premises. Under these circumstances, you should 3.____

   A. assure the owner that in the interest of good public relations the inspector involved will not be assigned to inspect the owner's premises in the future
   B. discuss the matter with the inspector before submitting your report
   C. inform the owner that complaints which cannot be substantiated cannot receive further consideration
   D. mark the complaint *not substantiated* and refrain from discussing it with the inspector

4. You are assigned to work with another inspector on a complex inspectional problem which has received considerable newspaper publicity. After the field work is completed, you agree to prepare a report incorporating both findings. You prepare the report and submit it directly to your superior without showing it to your fellow inspector. At a staff conference, your superior praises your report and the work performed by you; he minimizes the performance of your fellow inspector. You remain silent. Later you learn that your fellow inspector has been aggrieved by your conduct.
Of the following courses of action, it is MOST advisable that you 4.____

   A. ask the inspector whether he wishes you to try to get more credit for him
   B. discuss the matter with your fellow inspector and later with your superior to point out that the inspection was a joint effort and that your colleague should share in the credit
   C. ignore the matter and allow time to take care of the incident
   D. write a memorandum to your superior detailing precisely the work performed by the other inspector in connection with the assignment

5. Assume that you are a licensed pharmacist and would like to secure a part-time job as a pharmacist to supplement your income. In the course of your work as an inspector, you investigate an anonymous complaint against a drug store. Your investigation discloses nothing to indicate that the drug store owner has violated any provision of the Code. The owner, learning that you are a licensed pharmacist, asks you to work for him on a part-time basis. Under the circumstances, you SHOULD 5.____

   A. *accept* provided that the owner has not asked you for special consideration
   B. *accept* provided that you will receive the union scale of wages
   C. *refuse* the offer since a conflict of interest situation may be involved
   D. *refuse* the offer until you have a chance to discuss it with other inspectors who are licensed pharmacists

6. Your supervisor frequently bypasses you and assigns work directly to your subordinates. You had called this matter to his attention previously. At that time, he assured you that you would not be bypassed again. However, he has continued to bypass you.
Under these circumstances, you SHOULD 6.____

   A. attempt to determine the reasons for your supervisor's action before proceeding further
   B. begin keeping a record of the instances when you are bypassed, and forward a memorandum to your supervisor's superior setting forth such instances
   C. ignore the situation until such time as your supervisor brings the matter up for discussion
   D. instruct your staff that they are to accept assignments only from you

7. In checking the daily reports of one of your inspectors, you notice that he is consistently late in beginning his working day. You discuss the matter with him and point out that disciplinary action may be taken unless he starts work promptly. He denies that he is tardy in beginning his work day. However, based on your field follow-up visits, the evidence indicates that the inspector continues to be late in starting work. Again, you discuss the matter with him and he again denies your contention that he is late in starting work.
You should 7.____

A. again point out the need for starting work promptly and continue checking the inspector's starting time
B. discuss the matter with other inspectors in the work group to get their advice
C. ignore the matter as long as the inspector makes about as many inspections as others in the group
D. report the inspector to your supervisor for appropriate disciplinary action

8. One of the inspectors in your group makes about one-fourth more inspections than any other inspector. However, his inspections do not meet satisfactory standards of quality. After you have given him training in the field, his work improves to a point where it is satisfactory. However, he still makes about one-fourth more inspections than any of the other inspectors. 8.____
You should

A. ask that the inspector be transferred to a unit where the quantity and quality of work produced by him will be closer to the group standard
B. ask the proprietors of establishments visited by this inspector whether the inspections were too cursory
C. devote less time to this inspector so that you may devote more time to those inspectors who may need additional training
D. instruct the inspector to reduce the number of his inspections to the group standard and spend more time in each establishment

9. Assume that your superior has given you an additional job to do which will require extra effort on the part of your inspectors who are now carrying a full work load. You feel that the job cannot be completed in the allotted time. You present your point of view but your superior insists that you handle the assignment without any increase in staff. 9.____
Of the following courses of action, it would be MOST advisable for you to

A. attempt to complete the assignment within the allotted time by rescheduling and re-assigning other work
B. commit yourself to no specific course of action while attempting to secure evidence to support your position that you should not be given the assignment
C. insist that your superior give you some assurance that this assignment does not set a precedent for assignments of a similar nature and agree to do the job
D. take the matter up with higher authority, preferably by memorandum, but apprise your superior of your action

10. You are conducting a conference with the inspectors assigned to you. During the conference, you make a statement regarding field inspections which you are reasonably certain is correct. One of the inspectors tells you in an offensive manner that your statement is incorrect. Some of the inspectors agree with him; others remain silent. Under these circumstances, you SHOULD 10.____

A. ask the inspectors who have not made any comments for their opinions and be guided by their remarks
B. ignore the offensive manner of the speaker and state that since you are certain that you are correct, the group will be guided by your statement
C. state that while the manner of the speaker is offensive he is nevertheless probably correct
D. state that you will ascertain whether your statement is correct and will advise them of it in the near future

11. Assume that you are in the habit of writing to your supervisor on subjects related to your duties. Your supervisor tells you that you are writing too many memorandums to him. Of the following courses of action, it is MOST preferable for you to 11.____

A. instruct your inspectors not to put in writing communications regarding the work of the unit
B. refrain from communicating in writing with your supervisor
C. take no notice of your supervisor's statement since the smooth functioning of an organization depends upon written communication
D. write to your supervisor only when you feel that it is necessary

12. You are accompanying one of your recently appointed inspectors on a field inspection. His inspections take an unusually long time to complete since he is extremely meticulous. In these circumstances, you should FIRST 12.____

A. assure the inspector of your confidence in his ability to perform his job properly after sufficient training before criticizing his work performance
B. seek the transfer of the inspector to a position in the department which does not require contact with the public
C. tell the inspector that if he does not bring his work up to standard immediately, you will report him to your supervisor
D. urge the inspector to seek employment in a field not related to his present work

13. A rumor has started among the members of your staff to the effect that you will soon be leaving government service to take a position in private industry. You know that the rumor is untrue. You SHOULD 13.____

A. ask your staff not to discuss matters among themselves which relate to your own affairs
B. inform your staff that you do not intend to take a position in private industry
C. say nothing about the matter to your staff
D. tell your staff that you refuse to confirm or deny rumors concerning your employment prospects

14. You request an inspector to do something in a certain manner. The inspector asks you the reason for performing the operation in the manner suggested by you. You SHOULD 14.____

A. change the subject of your discussion
B. explain to the inspector that it is his job to carry out instructions - not to evaluate them
C. give the inspector the reason for your request
D. tell the inspector that if he thinks about the matter he will be able to determine the reason himself

15. You are conducting a conference with your staff. One of your inspectors seems completely disinterested in the discussion. To get this inspector to participate, you SHOULD 15.____

A. ask the inspector direct questions related to the subject being discussed
B. determine if there is any subject this inspector would like the group to discuss

C. ignore the situation until such time as the inspector shows interest
D. tell the inspector in a polite way to pay strict attention

16. You are conducting a conference with your staff and are having a great deal of difficulty with one of the inspectors who wants to do all of the talking. You have previously spoken privately to this inspector regarding his habit of *hogging the discussion*—to no avail. Under the circumstances, you SHOULD 16.____

A. ask the inspector to act as an auditor only during conferences
B. elicit discussion by direct questioning of other members of the staff
C. refrain from looking at the inspector when you ask a question; this will make it impossible for him to *get the floor*
D. tell the inspector to remain silent or to leave the group

17. You telephone one of your inspectors, assigning him to the central office for a period of two days to perform clerical duties. The inspector complains loudly, tells you that he dislikes clerical work and that he is being treated unfairly since there are inspectors in other boroughs who are assigned less frequently to clerical duties. You explain the situation as best you can but the inspector continues to object.
Under these circumstances, you SHOULD 17.____

A. ask the inspector to disregard the assignment pending your inquiry into practices followed in other boroughs in assigning personnel to clerical duties
B. promise the inspector that in the future you will do your best to give clerical assignments to people who do not voice objections to such assignments
C. tell the inspector to report for duty in accordance with your instructions
D. tell the inspector to take the matter up with your superior

18. In developing an on-the-job training program for inspectors, the FIRST thing which should be determined is 18.____

A. areas in which training is needed
B. how many inspectors are interested in training
C. how much will the training program cost
D. what training aids and facilities are available

19. Assume that a recently appointed and inexperienced inspector is given a difficult assignment. He is not given any specific instructions as to how the assignment should be carried out.
Such action is 19.____

A. *good;* a new employee needs to be encouraged to exercise his own initiative
B. *good;* a new employee will remember longer if he learns by himself
C. *poor;* newly appointed employees usually need guidance
D. *poor;* the cost of training varies from employee to employee

20. A number of important changes have taken place in several sections of the Code. You are to inform a group of inspectors of these changes and how they are to be implemented. While speaking to the group concerning the changes, one of the inspectors whom you know to be a quick learner complains that you are proceeding too slowly; another inspector whom you know to be the slowest learner in the group tells you that your teaching pace
is just right. You SHOULD 20.____

A. bring the session to a halt and instruct group members on an individual basis
B. proceed at a faster rate
C. proceed at a faster rate but allow more time for *breaks*
D. proceed at the same rate

21. As part of a new inspector's training, you observe him as he conducts an inspection. The inspector completes the *score-card* on which he lists certain violations of the Code. You look at the *score-card* and note that although the inspector spoke to the establishment owner about a certain violation, the inspector failed to list the violation on the *score-card.* Of the following, the MOST desirable way of pointing out this omission is to 21.____

A. ask the inspector to look at the *score-card* to see if anything is missing
B. criticize the inspector in a forthright manner and impress upon him the importance of the probationary period
C. show the *score-card* to the owner and ask the owner to indicate the violation which was noted but not recorded
D. tell the inspector in the presence of the owner to list the violation and make a separate note of the omission for service rating purposes

22. Assume that you are in the field training a recently appointed inspector in inspectional techniques.
The inspections demonstrated by you SHOULD be of the kind 22.____

A. consistent with the high standards of experienced inspectors
B. performed by the average beginning inspector so as not to unduly discourage the trainee
C. performed by the sanitarian who barely meets the minimum acceptable standard
D. which varies sharply from one inspection to the next so that the new inspector will be able to familiarize himself with various ways in which inspections may be conducted

23. The one of the following which LEAST describes the function of planning at the senior inspector level of supervision is deciding 23.____

A. *how* something should be done
B. *what* must be done
C. *who* should do it
D. *why* something should be done

24. For effective management, delegation of responsibility MUST be accompanied by *appropriate* 24.____

A. authority
B. commendation
C. compensation
D. privilege

25. The one of the following which is LEAST a *staff* function in an organization is 25.____

A. advising
B. directing
C. observing
D. planning

26. Assume that an employee is responsible to two supervisors of equal rank for the proper performance of his duties. The principle of good management which is NOT being complied with is 26.____

A. delegation of authority
B. fixed responsibility
C. homogeneous assignment
D. unity of command

27. Where low morale is responsible for low work output, the FIRST step which should be taken is to 27.____

A. determine the reason for the poor state of morale by interviewing supervisors and employees who are directly affected
B. have the head of the organization deliver an inspirational talk to those responsible for the low work output, stressing the mission of the organization and the importance of the work involved
C. lower standards of production to equal work output and then gradually increase these standards to the desired level
D. withdraw privileges with regard to the granting of leave, coffee breaks, and choice of lunch hours until work output rises to a satisfactory level

28. The one of the following which is NOT usually a need which gives rise to a work simplification program in government is the need to 28.____

A. make the job as pleasant as possible for employees
B. make things more convenient for members of the public
C. produce a greater quantity and higher quality of work
D. provide additional employment in times of recession

29. The MOST practical control the inspector has over the contractor when the inspector is not satisfied with the quality of the work is to 29.____

A. discuss withholding payment on that part of the work that is unsatisfactory
B. threaten to have the contractor thrown off the job
C. request that the contractor fire the men responsible for the unsatisfactory work
D. call the owner of the company and explain the situation to him

30. In the absence of a formal training program for inspectors, the BEST of the following ways to train a new man who is to do inspection work is to 30.____

A. give him the literature on the subject so that he can learn what he has to know
B. have him accompany an inspector as the inspector does his work so that he can learn by observing
C. assign him the job and let him learn on his own
D. tell him to go to a school at night that specializes in this field so that he will gain the necessary background

KEY (CORRECT ANSWERS)

1.	B	16.	B
2.	A	17.	C
3.	B	18.	A
4.	B	19.	C
5.	C	20.	D
6.	A	21.	A
7.	D	22.	A
8.	C	23.	D
9.	A	24.	A
10.	D	25.	B
11.	D	26.	D
12.	A	27.	A
13.	B	28.	D
14.	C	29.	A
15.	A	30.	B

TEST 2

DIRECTIONS: Each question or incomplete statement is followed by several suggested answers or completions. Select the one that BEST answers the question or completes the statement. *PRINT THE LETTER OF THE CORRECT ANSWER IN THE SPACE AT THE RIGHT.*

1. Assume that you are a supervisor newly assigned to a squad of inspectors. In order to establish a favorable working atmosphere, it is BEST to 1.____

   A. discipline ineffective members of your squad at regular intervals
   B. speak extensively on job-related subjects
   C. give advice on personal matters
   D. recognize and accept ideas submitted by members of your squad

2. Assume that you are a supervisor who has in his squad an ambitious inspector studying for promotion. This man takes every opportunity to ask you questions about your job. Under the circumstances, it is BEST for you to 2.____

   A. remind him firmly that he already has a full-time job and that if he wishes to study for promotion he should do it off-duty by himself
   B. plan your time so that you can assist in his promotional aspirations
   C. tell him that you would like to help but that you do not wish to give him an advantage over others
   D. resist instructing him because if he is promoted you will lose a valuable man, thereby weakening your squad

3. A signed written complaint has been mailed directly to you alleging that one of your inspectors has been overly aggressive in that he pushed the complainant. The inspector is a good worker, and this is the first complaint ever recorded against him. Under the circumstances, it is BEST to 3.____

   A. notify informally the accused inspector of the nature of the complaint, and suggest that he guard his behavior in the future
   B. ignore the complaint as being too vague to warrant action
   C. have the complainant carefully investigated to see whether he has made similar complaints in the past
   D. have the complaint investigated by someone disinterested in the outcome of the matter

4. Assume that you are a supervisor in charge of an inspector who has a good work record but who, for the first time, exhibits symptoms of drunkenness. When confronted, he denies that he ever drinks and says that his apparently intoxicated behavior is really the result of his doctor's medication for the flu. Under the circumstances, it is BEST to 4.____

   A. ignore the situation for the present but later report the matter to your superiors
   B. tell the man that you know he's untruthful but that, because of his previous good record, you are willing to overlook his condition this time
   C. accept the man's explanation, send him home for the day on sick leave, but watch for future symptoms of possible drunken behavior
   D. reprimand the man, send out for coffee to sober him up, and warn him the next time he exhibits drunken symptoms he will face severe disciplinary action

5. Assume that you are in charge of a squad of inspectors. One inspector has been performing ineffectively, although working hard. All attempts to improve his performance have failed. He is nearing the end of his probationary period. 5.____
In the circumstances, it is BEST to

   A. reschedule assignments so that the rest of the squad takes over a greater share of the work load
   B. recommend separation on the ground that improvement cannot be achieved
   C. assign only the simplest cases to the man
   D. leave the man alone, since he seems to be doing the best he can

6. As a supervisor, you have been instructed by your superiors to install a radically revised system of procedure for your squad. You are concerned that your subordinates may resist the change. 6.____
The BEST way for you to secure the willing cooperation of your squad in effecting the change is to

   A. secure the participation of all your subordinates in planning for the change, emphasizing the absence of any threat to their security
   B. *sell* your subordinates on the new procedure by emphasizing that the procedure has the full backing of your superiors
   C. warn your subordinates not to sabotage the change, emphasizing that willful interference with the change will be followed by severe corrective disciplinary action
   D. appeal to your subordinates' loyalty to the agency and to yourself, emphasizing that *one hand washes the other*

7. Supervising inspectors are involved in the decision-making process. 7.____
Effective decision-making means MOST NEARLY

   A. compromising, since all decisions involve compromise
   B. selecting the course of action with the least unexpected consequences
   C. holding off on any action until circumstances dictate one particular approach
   D. securing employee participation in the planning and policy process

8. Assume that you are a supervisor in charge of an inspector who may be abusing sick leave. 8.____
Under the circumstances, the FIRST thing you should do is to

   A. interview the inspector to find out what is wrong
   B. maintain a calendar of sick leave used by the inspector to see whether a pattern develops indicating abuse
   C. warn the inspector against any further malingering
   D. institute corrective disciplinary action the very next time the inspector reports sick

9. Supervision is a social relationship. It is both the art of being a leader and a subordinate. 9.____
This statement implies that

   A. the supervisory relationship involves an expectation of obedience on the part of the supervisor and a willingness to obey on the part of the subordinate
   B. the really successful supervisor always knows that his subordinates understand him, and doesn't have to clarify and explain his orders
   C. in the supervisory relationship, supervisor and subordinate should strive to be as friendly with each other as possible

D. the really wise subordinate knows his job and sees to it that his supervisor knows that he knows his job

10. As a supervising inspector, you have a man in your squad who avoids difficult tasks on the ground that he cannot do the more difficult work. You have informally condoned this practice because he is effective and busy on lesser tasks, overall squad production is satisfactory, and no one has complained. Nevertheless, you decide to review the situation. 10.____
Solely on the basis of the information presented, the LEAST effective response to this situation is to

A. denounce the man before the group and ask for their advice on handling the matter
B. insist on a basic work capability for all members of the squad
C. continue the present practice informally, so long as production and morale are unaffected
D. remind the man that professional recognition awaits those who work hard on a variety of tasks

11. Assume that you are a new supervisor in charge of a squad of inspectors. Your superior informs you that the squad has long been declining in effectiveness. Your job is to increase production without changing personnel. 11.____
Of the following, the MOST important information for you to have in order to effect change is

A. the reason for the squad's past production successes
B. an accurate account of your squad's present state of mind
C. a knowledge of the interplay of psychic needs and neighborhood surroundings in producing the squad's laxity
D. a case history on every individual so that you can estimate the personal impact of prospective changes

12. Assume that you have become the supervisor of a high morale squad of inspectors, all of whom are experienced and productive. 12.____
The BEST supervisory approach for you to take to insure the continuance of an efficient squad is to

A. *leave them alone,* since it doesn't pay to tinker with a well-running mechanism
B. develop a close personal relationship with the most experienced member of your squad and use this relationship to govern the rest of the squad
C. take charge immediately, and let them know who's in charge since everything usually runs well when persons are alert
D. work problems out together, on the theory that things usually run well when the supervisor successfully seeks to build power with, rather than hold authority over, his work group

13. One of the things a supervising inspector should AVOID doing is 13.____

A. answering unimportant questions asked by the public
B. talking to people he does not know
C. blaming his supervisors for all the unpleasant orders the supervising inspector must issue
D. showing an interest in public problems

14. An angry building owner complains loudly to you, the supervisor, about the actions of the inspectors assigned to you. 14.____
You should

A. try to find excuses for your men's actions
B. speak to him in the same tone of voice he is using
C. insist that the actions of your men are correct
D. try to answer his complaint quietly

15. In dealing with the general public, an inspector should remember that 15.____

A. every person is an individual who may think for himself
B. all people tend to think alike
C. most people think alike
D. it is best to change the public's way of thinking to what the department requires

16. An inspector is performing his job in the BEST manner when he 16.____

A. continually checks with his supervisor to make sure each inspection is being done properly
B. knows enough to overlook minor violations that have a negligible effect on overall
C. varies the rules when he feels they do not meet the conditions of the job
D. is careful and observant in his inspections

17. An IMPORTANT characteristic of a good supervisor is his ability to 17.____

A. be a stern disciplinarian
B. put off the settling of grievances
C. solve problems
D. find fault in individuals

18. At the time you hand out a job assignment, an inspector feels that he cannot complete the job within the time limit you have given him. 18.____
You would expect the inspector FIRST to

A. make as many inspections as possible and then report to you
B. compare his workload to that of the other inspectors
C. complete the work by putting in overtime before notifying you of the problem
D. request assistance in doing the work

19. A new supervising inspector will BEST obtain the respect of the men assigned to him if he 19.____

A. makes decisions rapidly and sticks to them regardless of whether they are right or wrong
B. makes decisions rapidly and then changes them just as rapidly if the decisions are wrong
C. does not make decisions unless he is absolutely sure that they are right
D. makes his decisions after considering carefully all available information

20. A newly-appointed inspector is operating at a level of performance below that of the other employees. 20.____
In this situation, a supervisor should FIRST

A. lower the acceptable standard for the new inspector
B. find out why the new inspector cannot do as well as the others
C. advise the new inspector he will be dropped from the payroll at the end of the probationary period
D. assign another new inspector to assist the first inspector

21. Assume that you have to instruct a new inspector on a specific departmental operation. The new man seems unsure of what you have said. 21.____
Of the following, the BEST way for you to determine whether the man has understood you is to

A. have the man explain the operation to you in his own words
B. repeat your explanation to him slowly
C. repeat your explanation to him, using simpler wording
D. emphasize the important parts of the operation to him

22. A supervising inspector realizes that he has taken an instantaneous dislike to a new inspector assigned to him. The BEST course of action for this supervisor to take in this case is to 22.____

A. be especially observant of the new inspector's actions
B. request that the new inspector be reassigned
C. make a special effort to be fair to the new inspector
D. ask to be transferred himself

23. A supervisor gives detailed instructions to his inspectors as to how a certain type of job is to be done. 23.____
One ADVANTAGE of this practice is that this will

A. result in a more flexible operation
B. standardize operations
C. encourage new men to learn
D. encourage initiative in the men

24. Of the following, the one that would MOST likely be the result of poor planning is: 24.____

A. Omissions are discovered after the work is completed
B. During the course of normal inspection, a meter is found to be unaccessible
C. An inspector completes his assignments for that day ahead of schedule
D. A problem arises during an inspection, and prevents an inspector from completing his day's assignments

25. Of the following, the BEST way for a supervisor to maintain good morale among his inspectors is for the supervisor to 25.____

A. avoid correcting an inspector when he makes mistakes
B. continually praise an inspector's work even when it is of average quality
C. show that he is willing to assist in solving the inspector's problems
D. accept the inspector's excuses for failure even though the excuses are not valid

KEY (CORRECT ANSWERS)

1. D
2. B
3. D
4. C
5. B
6. A
7. B
8. A
9. A
10. A
11. B
12. D
13. C
14. D
15. A
16. D
17. C
18. D
19. D
20. B
21. A
22. C
23. B
24. A
25. C

EXAMINATION SECTION
TEST 1

DIRECTIONS: Each question or incomplete statement is followed by several suggested answers or completions. Select the one that BEST answers the question or completes the statement. *PRINT THE LETTER OF THE CORRECT ANSWER IN THE SPACE AT THE RIGHT.*

1. A supervisor was given a booklet that showed a new work method that could save time. He didn't tell his men because he thought that they would get the booklet anyway. For the supervisor to have acted like this is a 1.____

   A. *good* idea because he saves the time and bother of talking to the men
   B. *bad* idea because he should make sure his men know about better work methods
   C. *good* idea because the men would rather read about it themselves
   D. *bad* idea because a supervisor should always show his men every memo he gets from higher authority

2. A supervisor found it necessary to discipline two subordinates. One man had been operating his equipment in a wrong way, while the other man came to work late for three days in a row. The supervisor decided to talk to both men together. For the supervisor to deal with the problems in this way is a 2.____

   A. *good* idea because each man will learn about the difficulties of the other person and how to solve such difficulties
   B. *bad* idea because the supervisor should wait until he can bring a larger group together and save time in discussing such questions
   C. *good* idea because he will be able to get the men to see that their problems are related
   D. *bad* idea because he should meet with each man separately and give him his full attention

3. A supervisor should try to make his men feel their jobs are important in order to 3.____

   A. get the men to say good things about their supervisor to his own superior
   B. get the men to think in terms of advancing to better jobs
   C. let higher management in the agency know that the supervisor is efficient
   D. help the men to be able to work more efficiently and enthusiastically

4. A supervisor should know approximately how long it takes to do a particular kind of job CHIEFLY because he 4.____

   A. will know how much time to take if he has to do it himself
   B. will be able to tell his men to do it even faster
   C. can judge the performance of the person doing the job
   D. can retrain experienced employees in better work habits

5. Supervisors often get their employees' opinions about better work methods because 5.____

   A. the men will know that they are respected
   B. the men would otherwise lose all their confidence in the supervisor
   C. the supervisor might find in this way a good suggestion he could use
   D. this is the best method for improvement of work methods

6. Right after you have trained your subordinates in doing a new job, you find that they seem to be doing all right, but that it will take them several days to finish. You also have several groups of men working at other locations. The MOST efficient way for you to make sure that the men continue doing the new job properly is to 6.____

   A. stay on that job with the men until it is finished just in case trouble develops
   B. visit the men every half hour until the job is done
   C. stay away from their job that day and visit the men the next day to ask them if they had any problems
   D. visit the men a few times each day until they finish the new job

7. Assume that one of your new employees is older than you are. You also think that he may be hard to get along with because he is older than you. The BEST way for you to avoid any problems with the older worker is for you to 7.____

   A. lay down the law immediately and tell the man he better not cause you any trouble
   B. treat the man just the way you would any other worker
   C. always ask the older worker for advice in the presence of all the men
   D. ignore the man entirely until he realizes that you are the boss

8. Assume that you have tried a new method suggested by one of your employees and find that it is easier and cheaper than the method you had been using.
The proper thing for you to do NEXT is to 8.____

   A. say nothing to anyone but train your men to use the new method
   B. train your men to use the new method and tell your crew that you got the idea from one of the men
   C. continue using the old method because a supervisor should not use suggestions of his men
   D. have your crew learn the new method and take credit for the idea since you are the boss

9. Suppose you are a supervisor and your superior tells you that the way your men are doing a certain procedure is wrong and that you should re-train your men as soon as possible.
When you begin to re-train the men, the FIRST thing you should do is 9.____

   A. tell your men that a wrong procedure had been used and that a new method must be learned as a result
   B. train your employees in the new method with no explanation since you are the boss
   C. tell the crew that your superior has just decided that everyone should learn a new method
   D. tell the crew that your superior says your method is wrong but that you don't agree with this

10. It is BAD practice to criticize a man in front of the other men because 10.____

   A. people will think you are too strict
   B. it is annoying to anyone who walks by
   C. it is embarrassing to the man concerned
   D. it will antagonize the other men

11. A supervisor decides not to put his two best men on a work detail because he knows that they won't like it. 11.____
For the supervisor to make the work assignment this way is a

A. *good* idea because it is only fair to give your best men a break once in a while
B. *bad* idea because you should treat all of your men fairly and not show favoritism
C. *good* idea because you save the strength of these men for another job
D. *bad* idea because more of the men should be exempted from the assignment

12. Suppose you are a supervisor and you find it inconvenient to obey an established procedure set by your agency. You think another procedure would be better. 12.____
The BEST thing to do first about this procedure that you don't like is for you to

A. obey the procedure even if you don't want to and suggest your idea to your own supervisor
B. disregard the procedure because a supervisor is supposed to have some privileges
C. follow the procedure some of the time but ignore it when the men are not watching
D. organize a group of other supervisors to get the procedure changed

13. A supervisor estimated that it would take his crew one workday per week to do a certain job each week. However, after a month he noticed that the job averaged two and a half days a week and this delayed other jobs that had to be done. 13.____
The FIRST thing that the supervisor should do in this case is to

A. call his men together and warn them that they will get a poor work evaluation if they do not work harder
B. talk to each man personally, asking him to work harder on the job
C. go back and study the maintenance job by himself to see if more men should be assigned to the job
D. write his boss a report describing in detail how much time it is taking the men to do the job

14. An employee complains to you that some of his work assignments are too difficult to do alone. 14.____
Which of the following is the BEST way for you to handle this complaint?

A. Go with him to see exactly what he does and why he finds it so difficult.
B. Politely tell the man that he has to do the job or be brought up on charges.
C. Tell the man to send his complaint to the head of your agency.
D. Sympathize with the man and give him easier jobs.

15. The BEST way for a supervisor to keep control of his work assignments is to 15.____

A. ask the men to report to him immediately when their jobs are finished
B. walk around the buildings once a week and get a firsthand view of what is being done
C. keep his ears open for problems and complaints, but leave the men alone to do the work
D. write up a work schedule and check it periodically against the actual work done

16. A supervisor made a work schedule for his men. At the bottom of it, he wrote, *No changes or exceptions will be made in this schedule for any reason.* 16.____
For the supervisor to have made this statement is

A. *good* because the men will respect the supervisor for his attitude
B. *bad because* there are emergencies and special situations that occur
C. *good* because each man will know exactly what is expected of him
D. *bad* because the men should expect that no changes will ever be made in the work schedule without written permission

17. Which one of the following would NOT be a result of a well-planned work schedule? 17.____
The schedule

A. makes efficient use of the time of the staff
B. acts as a check list for an important job that might be left out
C. will give an idea of the work to a substitute supervisor
D. shows at a glance who the best men are

18. A new piece of equipment you have ordered is delivered. You are familiar with it, but the men under you who will use it do not know the equipment. Of the following methods, which is the BEST to take in explaining to them how to operate this equipment? 18.____

A. Ask the men to watch other crews using the equipment.
B. Show one reliable man how to operate the equipment and ask him to teach the other men.
C. Ask the men to read the instructions in the manual for the equipment.
D. Call the men together and show them how to operate the equipment.

19. One supervisor assigns work to his men by calling his crew together each week and describing what has to be done that week. He then tells them to arrange individual assignments among themselves and to work as a team during the week. 19.____
This method of scheduling work is a

A. *good* idea because this guarantees that the men will work together
B. *bad* idea because responsibility for doing the job is poorly fixed
C. *good* idea because the men will finish the job in less time, working together
D. *bad* idea because the supervisor should always stay with his men

20. Suppose that an employee came to his supervisor with a problem concerning his assignment. 20.____
For the supervisor to listen to this problem is a

A. *good* idea because a supervisor should always take time off to talk when one of his men wants to talk
B. *bad* idea because the supervisor should not be bothered during the work day
C. *good* idea because it is the job of the supervisor to deal with problems of job assignment
D. *bad* idea because the employee could start annoying the supervisor with all sorts of problems

21. Suppose that on the previous afternoon you were looking for an experienced employee in order to give him an emergency job and he was missing from his job location. The next morning, he tells you that he got sick suddenly and had to go home, but could not tell you since you were not around. He has never done this before. 21.____
What should you do?

A. Tell the man he is excused and that in such circumstances he did the wisest thing.
B. Bring the man up on charges because whatever he says he could still have notified you.
C. Have the man examined by a doctor to see if he really was sick the day before.
D. Explain to the man that he should make every effort to tell you or to get a message to you if he must leave.

22. An employee had a grievance and went to his supervisor about it. The employee was not satisfied with the way the supervisor tried to help him and told him so. Yet, the supervisor had done everything he could under the circumstances. 22.____
The PROPER action for the supervisor to take at this time is to

A. politely tell the employee that there is nothing more for the supervisor to do about the problem
B. let the employee know how he can bring his complaint to a higher authority
C. tell the employee that he must solve the problem on his own since he did not want to follow the supervisor's advice
D. suggest to the employee that he ask another supervisor for assistance

23. In which of the following situations is it BEST to give your men spoken rather than written orders? 23.____

A. You want your men to have a record of the instructions.
B. Spoken instructions are less likely to be forgotten.
C. An emergency situation has arisen in which there is no time to write up instructions.
D. There are instructions on time and leave regulations which are complicated.

24. One of your employees tells you that a week ago he had a small accident on the job but he did not bother telling you because he was able to continue working. 24.____
For the employee not to have told his supervisor about the accident was

A. *good* because the accident was a small one
B. *bad* because all accidents should be reported, no matter how small
C. *good* because the supervisor should be bothered only for important matters
D. *bad* because having an accident is one way to get excused for the day

25. For a supervisor to deal with each of his subordinates in exactly the same manner is 25.____

A. *poor* because each man presents a different problem and there is no one way of handling all problems
B. *good* because once a problem is handled with one man, he can handle another man with the same problem
C. *poor* because the men will resent it if they are not handled each in a better way than others
D. *good* because this assures fair and impartial treatment of each subordinate

KEY (CORRECT ANSWERS)

1. B
2. D
3. D
4. C
5. C
6. D
7. B
8. B
9. A
10. C
11. B
12. A
13. C
14. A
15. D
16. B
17. D
18. D
19. B
20. C
21. D
22. B
23. C
24. B
25. A

TEST 2

DIRECTIONS: Each question or incomplete statement is followed by several suggested answers or completions. Select the one that BEST answers the question or completes the statement. *PRINT THE LETTER OF THE CORRECT ANSWER IN THE SPACE AT THE RIGHT.*

1. Jim Johnson has been on your staff for over four years. He has always been a conscientious and productive worker. About a month ago, his wife died; and since that time, his work performance has been very poor. 1.____
As his supervisor, which one of the following is the BEST way for you to deal with this situation?

    A. Allow Jim as much time as he needs to overcome his grief and hope that his work performance improves.
    B. Meet with Jim to discuss ways to improve his performance.
    C. Tell Jim directly that you are more concerned with his work performance than with his personal problem.
    D. Prepare disciplinary action on Jim as soon as possible.

2. You are responsible for the overall operation of a storehouse which is divided into two sections. Each section has its own supervisor. You have decided to make several complex changes in the storekeeping procedures which will affect both sections. 2.____
Of the following, the BEST way to make sure that these changes are understood by the two supervisors is for you to

    A. meet with both supervisors to discuss the changes
    B. issue a memorandum to each supervisor explaining the changes
    C. post the changes where the supervisors are sure to see them
    D. instruct one supervisor to explain the changes to the other supervisor

3. You have called a meeting of all your subordinates to tell them what has to be done on a new project in which they will all be involved. Several times during the meeting, you ask if there are any questions about what you have told them. 3.____
Of the following, to ask the subordinates whether there are any questions during the meeting can BEST be described as

    A. *inadvisable* because it interferes with their learning about the new project
    B. *advisable* because you will find out what they don't understand and have a chance to clear up any problems they may have
    C. *inadvisable* because it makes the meeting too long and causes the subordinates to lose interest in the new project
    D. *advisable* because it gives you a chance to learn which of your subordinates are paying attention to what you say

4. As a supervisor, you are responsible for seeing to it that absenteeism does not become a problem among your subordinates. 4.____
Which one of the following is NOT an acceptable way of controlling the problem of excessive absences?

    A. Distribute a written statement to your staff on the policies regarding absenteeism in your organization.
    B. Arrange for workers who have the fewest absences to talk to those workers who have the most absences.
    C. Let your subordinates know that a record is being kept of all absences.
    D. Arrange for counseling of those employees who are frequently absent.

5. One of your supervisors has been an excellent worker for the past two years. There are no promotion opportunities for this worker in the foreseeable future. Due to the city's present budget crisis, a salary increase is not possible.
Under the circumstances, which one of the following actions on your part would be MOST likely to continue to motivate this worker? 5.____

A. Tell the worker that times are bad all over and jobs are hard to find.
B. Give the worker less work and easier assignments.
C. Tell the worker to try to look for a better paying job elsewhere.
D. Seek the worker's advice often and show that the suggestions provided are appreciated.

6. As a supervisor in a warehouse, it is important that you use your available work force to its fullest potential. Which one of the following actions on your part is MOST likely to increase the effectiveness of your work force? 6.____

A. Assigning more workers to a job than the number actually needed.
B. Eliminating all job training to allow more time for work output.
C. Using your best workers on jobs that average workers can do.
D. Making sure that all materials and equipment used are maintained in good working order.

7. You learn that your storage area will soon be undergoing changes which will affect the work of your subordinates. You decide not to tell your subordinates about what is to happen.
Of the following, your action can BEST be described as 7.____

A. wise because your subordinates will learn of the changes for themselves
B. unwise because your subordinates should be advised about what is to happen
C. wise because it is better for your subordinates to continue working without being disturbed by such news
D. unwise because the work of your subordinates will gradually slow down

8. In making plans for the operation of your unit, you are MOST likely to see these plans carried out successfully if you 8.____

A. allow your staff to participate in developing these plans
B. do not spend any time on the minor details of these plans
C. base these plans on the past experiences of others
D. allow these plans to interact with outside activities in other units

9. As a supervisor in charge of the total operation of a food supply warehouse, you find vandalism to be a potentially serious problem. On occasion, trespassers have gained entrance into the facility by climbing over an unprotected 8-foot fence surrounding the warehouse whose dimensions measure 100 feet by 100 feet.
Assuming that all of the following would be equally effective ways in preventing these breaches in security in the situation described above, which one would be LEAST costly? 9.____

A. Using two trained guard dogs to roam freely throughout the facility at night.
B. Hiring a security guard to patrol the facility after working hours.
C. Installing tape razor wire on top of the fence surrounding the facility.
D. Installing an electronic burglar alarm system requiring the installation of a new fence.

10. The area for which you have program responsibility has undergone recent changes. Your staff is now required to perform many new tasks, and morale is low. 10.____
The LEAST effective way for you to improve long-term staff morale would be to

A. develop support groups to discuss problems
B. involve staff in job development
C. maintain a comfortable social environment within the group
D. adequately plan and give assignments in a timely manner

11. As a supervisor in a large office, one of your subordinate supervisors stops you in the middle of the office and complains loudly that he is being treated unfairly. The rest of the staff ceases work and listens to the complaint. The MOST appropriate action for you to take in this situation is to 11.____

A. ignore this unprofessional behavior and continue on your way
B. tell the supervisor that his behavior is unprofessional and he should learn how to conduct himself
C. explain to the supervisor why you believe he is not being treated unfairly
D. ask the supervisor to come to your office at a specific time to discuss the matter

12. You are told that one of your subordinates is distributing literature which attempts to recruit individuals to join a particular organization. Several workers complain that their rights are being violated. Of the following, the BEST action for you to take FIRST is to 12.____

A. ignore the situation because no harm is being done
B. discuss the matter further with your supervisor
C. ask the worker to stop distributing the literature
D. tell the workers that they do not have to read the material

13. You have been assigned to develop a short training course for a recently issued procedure. 13.____
In designing this course, which of the following statements is the LEAST important for you to consider?

A. The learning experience must be interesting and meaningful in terms of the staff member's job.
B. The method of teaching must be strictly followed in order to develop successful learning experiences.
C. The course content should incorporate the rules and regulations of the agency.
D. The procedure should be consistent with the agency's objectives.

14. As a supervisor, there are several newly-promoted employees under your supervision. Each of these employees is subject to a probationary period PRIMARILY to 14.____

A. assess the employee's performance to see if the employee should be retained or removed from the position
B. give the employee the option to return to his former employment if the employee is unhappy in the new position
C. give the employee an opportunity to learn the duties and responsibilities of the position
D. judge the employee's potential for upward mobility in the future

15. An employee under your supervision rushes into your office to tell you he has just received a telephone bomb threat. 15.____
As the administrative supervisor, the FIRST thing you should do is

A. evacuate staff from the floor
B. call the police and building security
C. advise your administrator
D. do a preliminary search

16. After reviewing the Absence Control form for a unit under your supervision, you find that one of your staff members has a fifth undocumented sick leave within a six-month period. 16.____
In this situation, the FIRST action you should take is to

A. discuss the seriousness of the matter with the staff member when he returns to work and fully document the details of the discussion
B. review the case with the location director and warn the staff member that future use of sick leave will be punished
C. submit the proper disciplinary forms to ensure that the staff member is penalized for excessive absences
D. request that the timekeeper put the staff member on doctor's note restriction

17. A subordinate supervisor recently assigned to your office begins his first conference with you by saying that he has learned something that another supervisor is doing that you should know about. 17.____
After hearing this statement, of the following, the BEST approach for you to take is to

A. explain to the supervisor that the conference is to discuss his work and not that of his co-workers
B. tell the supervisor that you do not encourage a spy *system* among the staff you supervise
C. tell the supervisor that you will listen to his report only if the other supervisor is present
D. allow the supervisor to continue talking until you have enough information to make a decision on how best to respond

18. Assume that you are a supervisor recently assigned to a new unit. You notice that, for the past few days, one of the employees in your unit whose work is about average has been stopping work at about four o'clock and has been spending the rest of the afternoon relaxing at his desk. 18.____
The BEST of the following actions for you to take in this situation is to

A. assign more work to this employee since it is apparent that he does not have enough work to keep him busy
B. observe the employee's conduct more closely for about ten days before taking any more positive action
C. discuss the matter with the employee, pointing out to him how he can use the extra hour daily to raise the level of his job performance
D. question the previous supervisor in charge of the unit in order to determine whether he had sanctioned such conduct when he supervised that unit

19. A new supervisor was assigned to your program four months ago. Although he tries hard, he has been unable to meet certain standards because he still has a lot to learn. As his supervisor, you are required to submit performance evaluations within a few days. How would you rate this employee on the tasks where he fails to meet standards because of lack of experience? 19.____

A. Satisfactory
B. Conditional
C. Unsatisfactory
D. Unratable

20. You find that there is an important procedural error in a memo which you distributed to your staff several days ago. The BEST approach for you to take at this time is to 20.____

A. send a corrected memo to the staff, indicating what prior error was made
B. send a corrected memo to the staff without mentioning the prior error
C. tell the staff about the error at the next monthly staff meeting
D. place the corrected memo on the office bulletin board

21. Your superior asks you, a supervisor, about the status of the response to a letter from a public official concerning a client's case. When you ask the subordinate who was assigned to prepare the response to give you the letter, the subordinate denies that it was given to him. You are certain that the subordinate has the letter, but is withholding it because the response has not yet been prepared.
Of the following, in order to secure the letter from the subordinate, you should FIRST 21.____

A. accuse the subordinate of lying and demand that the letter be given to you immediately
B. say that you would consider it a personal favor if the subordinate would find the letter
C. continue to question the subordinate until he admits to having been given the letter
D. offer a face-saving solution, such as asking the subordinate to look again for the letter

22. As a supervisor, you have been assigned to write a few paragraphs to be included in the agency's annual report, describing a public service agency department this year as compared to last year.
Which of the following elements basic to the agency is LEAST likely to have changed since last year? 22.____

A. Mission
B. Structure
C. Technology
D. Personnel

23. As a supervisor, you have been informed that a grievance has been filed against you, accusing you of assigning a subordinate to out-of-title tasks. Of the following, the BEST approach for you to take is to 23.____

A. waive the grievance so that it will proceed to a Step II hearing
B. immediately change the subordinate's assignment to avoid future problems
C. respond to the grievance, giving appropriate reasons for the assignment
D. review the job description to ensure that the subordinate's tasks are not out-of-title

24. Which of the following is NOT a correct statement about agency group training programs in a public service agency? 24.____

A. Training sessions continue for an indefinite period of time.
B. Group training sessions are planned for designated personnel.
C. Training groups are organized formally through administrative planning.
D. Group training is task-centered and aimed toward accomplishing specific educational goals.

25. As a supervisor, you have submitted a memo to your superior requesting a conference to discuss the performance of a manager under your supervision. The memo states that the manager has a good working relationship with her staff; however, she tends to interpret agency policy too liberally and shows poor administrative skills by missing some deadlines and not keeping proper controls. 25.____
Which of the following steps should NOT be taken in order to prepare for this conference with your superior?

A. Collect and review all your notes regarding the manager's prior performance.
B. Outline your agenda so that you will have sufficient time to discuss the situation.
C. Tell the manager that you will be discussing her performance with your superior.
D. Clearly define objectives which will focus on improving the manager's performance.

KEY (CORRECT ANSWERS)

1. B
2. A
3. B
4. B
5. D

6. D
7. B
8. A
9. C
10. C

11. D
12. C
13. B
14. A
15. B

16. A
17. D
18. C
19. B
20. A

21. D
22. A
23. C
24. A
25. C

Control of Rodents and Insects

CONTROL OF RODENT INFESTATION AND HORBORAGE

In combating rodent infestation the use of cats, traps and poisons are only temporary expendients and do not eliminate rodent life completely from your premises. The best method of permanently eliminating them is to *build the out.* Rodent life exists in buildings because of favorable conditions that permit them to hide, nest and breed. They will not remain where safe shelter or food is not available. To combat infestation in your premises, it is necessary to be able to recognize rodent harborages or hiding places, both actual and potential as they are the conditions favoring rodent life and propagation. There are three general types of rodent harborages:

1- Temporary
2- Incidental
3- Structural

TEMPORARY RODENT HARBORAGES

These are conditions arising out of failure to maintain premises in a clean and sanitary condition, or faulty methods of operation, housekeeping, or storage of stock.

EXAMPLES

I. Mass storage of office supplies and old records, materials for repairs, food products or other store merchandise; boxes, crates, or cartons that are left undisturbed for periods of time and not rotated in use (using up older stock first).

II. Unused or obsolete fixtures or equipment, especially those having drawers, compartments or other hollow enclosures.

III. Miscellaneous junk, trash, odds and ends placed in closets, cellars, boiler rooms or out-of-the-way places, or portions of premises not in daily use having very little or no light.

IV. Garbage cans left uncovered overnight or having poorly fitting covers, or in a defective leaking condition.

V. Passageways used in transporting or storing garbage cans for removal, with spilled particles of food on floors, especially in corners.

VI. Accumulations of rubbish at botton of airshafts, dumbwaiter or elevator shaft pits, under sidewalk or cellar window gratings, or other parts of premises not cleaned regularly.

PREVENTION

I. Unused materials should be stored neatly and away from walls, allowing enough space for a man to pass around in cleaning and should preferably be stored sufficiently high above the floor to permit cleaning underneath. The amount stored should be minimized as much as possible, and it should be disturbed or its position changed at least every three weeks to prevent nesting of rodents.

II. Avoid mass storage by arranging in rows with 2' wide aisles. If stock is placed on shelves, raise the lowest shelf about 6" to 8" above the floor. Remove all rubbish that usually accumulates about unused materials. Promptly clean up food scrapes that spill from garbage cans, or fall under, or behind slop sinks, equipment, and stock bins. (Rodents feed more readily on these than on bagged or packaged food supplies). Store all garbage in non-leaking metal receptacles with tightfitting covers.

III. Place soiled linen into suitable containers.

IV. Maintain clean and sanitary conditions at all times.

INCIDENTAL RODENT HARBORAGES

These are conditions arising from installing of fixtures or equipment incidental to their use on the premises, in such a manner that hollow spaces, enclosures, and inaccessible places are formed.

EXAMPLES

I. Fixtures, refrigerators, ovens, etc. not installed flush against walls but leaving a small space that is too narrow for proper inspection and cleaning.

II. Narrow spaces left between bottoms of counters, back bars, or other fixtures or equipment, and the floor.

III. Small spaces existing between ceilings and tops of fixtures, clothes lockers, refrigerators, closets and cabinets, large overhead pipes and ventilating ducts suspended a few inches from ceiling.

IV. Hollow partitions (double wall space).
Hollow furniture of fixtures with inaccessible enclosures. Boxed-in casings or sheathing around pillars, pipes, radiators, etc., forming hollow enclosures.

V. Bottom shelves, stock platforms or skids that are not set directly on the floor but allow a space of a few inches to exist underneath.

VI. Defective insulated sections of large refrigerators or pipe coverings (hollow enclosed spaces formed by damage to cork or asbestos).

VII. Loose foods stored in low, thin, wooden food bins, boxes, cartons, burlap bags, etc.

VIII. Partially enclosed spaces behind open metal grilles used on housing of motors or other mechanical equipment.

PREVENTION

I. Eliminate narrow, inaccessible spaces behind fixtures or equipment by placing flush against wall or leaving a space wide enough for inspection and cleaning. Solidly block out narrow spaces underneath, or install flush on floors or raise high enough for cleaning.

II. Avoid providing undisturbed rat runways in narrow space between ducts or long hoods, and the ceiling. Ducts should be placed flush against ceilings and preferably be found in shape, instead of square.

III. Remove decorative boxing-in around radiators, columns, etc., to avoid hollow enclosures, or protect gnawing margins with metal flashing extending at least 6" above the floors, if they must be sheathed for appearance, used sheet metal.

IV. Repair and securely close all breaks in insulation around pipes, refrigerators or cooling cabinets.

V. Line interiors of wooden bins with sheet metal, or store foods in rodent proof screening (mesh openings not greater than 1/4").

VI. Eliminate hollow spaces formed by false bottoms in counters, lockers, cabinets, back bars, etc.

VII. Alter hollow fixtures so that enclosures are exposed for easy cleaning.

STRUCTURAL RODENT HARBORAGES

These conditions are due to design or construction of a building that are defective from a rat-proof standpoint, or that developed during occupancy from failure to make proper repairs or to use rat-proof materials.

EXAMPLES

I. Openings made in outside building walls, around beams, or in interior walls, floors or ceiling for installation of pipes, cables, or conduits. They are made by plumbers, electricians, or other workmen. The openings are usually larger than necessary and the unused portions of holes are not closed up. Holes, large cracks, loose bricks, or other openings in floors, walls or ceilings are other examples.

II. Hollow spaces in double walls, between floor and ceiling of lower story, and in double ceilings of cellars.

III. Enclosed hollow spaces formed by sheathing the undersides of stairways, by installation of false floors in toilets, or by raised wooden floors over earthen floors of cellars.

IV. Entrance and cellar doors that are not tight-fitting or not provided with a proper door sill or saddle, permitting openings over 1/4" to exist and not protected around gnawing edges with metal flashing at least 6" above floor level.

V. Openings around ceiling or floor beams, or risers, where they pass through partitions.

VI. Openings of fans, ventilators, and louvers on the outside of buildings, or fancy metal grilles with openings over 1/4", not protected by rodent proof screening.

VII. Floor drain and sewer trap pits not kept clean and not provided with solid metal covers with preforations not exceeding 1/4". Cellar floors of earth, enabling rodents to burrow underneath.

PREVENTION

I. Promptly seal up all holes or openings around pipe lines or cables where they enter the building, with concrete mortar or cement mortar to which ground glass may be added for better results.

II. Place tight-fitting metal collars or flanges around pipes and risers. Provide escutcheon plates for all risers where they pass through floor slabs, unless same are waterproofed by pockets of mastic.

III. Seal up all openings around beams.

IV. Avoid using double-wall type construction with hollow interior spaces, or hollow tile block, hollow cement block, or similar material for partitions or walls of storage compartments or in cellars.

V. Inspect all parts of premises for holes and seal every opening in walls and ceilings with cement plastered smooth. Move away fixtures and stock that may hide holes in floors and use a flashlight so as not to miss any. Look for loose bricks, cracks or other openings in cellar foundation walls. Find all openings before rodents do. Inspect regularly and repair weak spots before actual breaks occur.

VI. Block out hollow spaces under raised wooden floors with concrete Refrigerators, ranges, ovens, etc., should be solidly based on concrete. Protect entrance, cellar doors, and windows with metal flashing around gnawing edges, and maintain in good repair.

VII. Replace earthen cellar floors with a floor of concrete at least 3" - 4" thick and tied securely into foundation walls.

VIII. Securely anchor window and door screens to the frames.

RODENT INFESTATION SURVEYS

In addition to trapping, surveys by operators will indicate presence and approximate extent of infestation, of which the following are some signs.

Excreta of Pellets

Physical state will indicate recent or old infestation. Soft, moist droppings indicate live rats or mice present, while hard and dry ones indicate old. Amount of droppings indicate heavy or light infestation. The size of pellets will show if the rodents are large or small; and if different sizes are present, it indicates litters of young are being reared.

Gnawings

If recent, will show fresh appearance of gnawings, shavings, debris or marks on food bags or containers, or damage to other merchandise, supplies or fixtures.

Rat Run

Difficult to tell by appearance if new or old. Use white chalk or paint on suspected rat run. The rat is a creature of habit and will continue to use the same pipe or beam. It will leave marks caused by dirt or grease on feet or fur.

FOR PERMANENT CONTROL MEASURES

I. Try to maintain permanent freedom rather than resort to temporary reduction of rat population by periodic drives employing trapping, poisoning or fumigation.

II. After carrying out all rat stoppage measures, a reliable employee should inspect entire premises weekly to insure cleaning and upkeep, and to repair any temporary breakages in windows and doors to the outside. Allow no accumulations of rubbish to form. Watch sky-lights, air shafts, dumbwaiter and elevator shafts, and all other means of ingress from outside, for breaks. Immediate repairs to be made to eliminate openings and harborages, with rat-proof material (impervious to gnawing). Relocate or alter fixtures with hollow enclosures. Prevent careless employees or workmen from leaving lower windows or cellar doors overnight or on weekends to rodent ingress from outside sources.

III. Include in the specifications for all new construction and repair contracts a specific provision that work is be done is to leave the building in a ratproof condition. Specifications may read as follows:

> *"This building is planned and detailed, and it is the intent of these specifications, to provide a structure that will prevent the penetration by rodent vermin of any vacant space where they might find a harborage. The contractor will he held responsible for securing this condition by the closing of all points of access to such spaces, including the passage of piping and conduits through all walls, partitions, ceilings and furred off spaces, the closing of access to void in hollow tile blocks, etc. There shall be a special inspection of the building with regard to this matter before final acceptance."*

IV. All permanent measures are aimed at eliminating the rodent's food supply and shelter. Architects need to be made more cognizant of conditions that prevent rodent harborage and infestation, so as to change design of new buildings to eliminate unnecessary enclosed spaces. Rat-proof construction should receive greater prominence in the future.

INSECT CONTROL

Proper restaurant or food plant sanitation must obviously include measures for the elimination of insects and vermin. Where insects infest food, such food cannot be used for human consumption and must be destroyed, entailing an economic loss. Even more important is the fact that insects carry millions of bacteria in and on their bodies and contaminate the food on which they crawl. Food establishments are particularly susceptible to insect infestation are attracted by the food.

GENERAL ENTOMOLOGY

Life Cycle: Insects have a life cycle that consists of: (a) the egg, (b) the larva (worm-like), (c) the pupa (web, cocoon) and (d) the adult. Some insects have only three stages: (a) egg, (b) nymph, and (c) adult.

Structure: The general external structure of insects consists of three parts:

1. Head

   (a) Compound eyes - often more efficient than human eyes; enables them to see in all directions.
   (b) Antennae - hair-like structures that act as feelers.
   (c) Mouth - may be constructed for use as a hypodermic needle (mosquito), for chewing (beetle) or sucking (fly).

2. Thorax

   (a) Wings
   (b) Spiracles (breathing vents)
   (c) Feet

3. Abdomen - used for digestion, reproduction, and excretion. Insect "specks" may be from excreta or vomit.

Flies are vicious public health enemies. They breed in filth and carry diarrheal diseases to the food we eat. Flies have a life span of about four weeks. One female may lay 150 eggs at one time and usually prefers manure, garbage, or some other filthy waste matter for this purpose. In four weeks, the eggs hatch in eight hours into the larva stage (maggots) which become fully developed in four to five days. Each larva then becomes a pupa and five days later the adult fly emerges. Hence there are ten days from the egg to adult stage but this elapsed time varies according to the temperature. The female may start laying eggs four days after reaching the adult stage.

Flies generally die in the early autumn, partly due to a fungus disease that attacks them, and partly due to the cold temperature. They do not hibernate, but larvae may survive in manure piles or beneath the soil over the winter months and start the new generations of flies the following summer.

The fly cannot eat solid food. It alights on food and vomits a liquid through its proboscis-like mouth. The dissolved food is then sucked back, but, of course, all or a portion of this "vomit" may remain on the food. The excreta also remains on the food.

Examination of flies has disclosed as many as 28,000,000 bacteria on the inside and 5,000,000 on the outside of the body of the fly. Flies infest manure piles, privies, toilets, and other waste materials and then may alight on food. It is obvious that they may easily pick up millions of harmful bacteria and infect our food. They constitute a dangerous channel of disease transmission that must be eliminated.

A very large outbreak of bacillary dysentery in an army encampment was definitely traced to flies as the vectors of this disease. Thousands of cases of illnesses are undoubtedly due to flies, even though it may be difficult to prove.

Prevention: Remove all breeding places that may be found in or adjacent to the establishment, e.g., piles of manure, garbage, filth in general. Keep garbage cans tightly covered, and thoroughly clean them when emptied. Screen all openings to the outer air.

Destruction: Swat the fly, use fly paper, or use fly sprays.

Fly Sprays: These generally contain a contact poison, such as pyrethrum, as the active ingredient, dissolved in a nonodorous organic solvent, usually kerosene which has been deodorized or to which some perfume has been added. Such sprays act on the nervous system of the fly and are very effective, provided the spray contacts the insect.

Pyrethrum: Is a powder that is obtained from the chrysanthemum flower. The active ingredients are known as pyrethrins and a good insect spray should contain 100-120 milligrams of pyrethrins per 100 cubic centimeters of liquid.

Pyrethrum is not toxic to man but kills the fly by contact. When a dose of this insecticide is absorbed, it stupefies the insect by acting on the nervous system and causes the insect to drop to the floor. It may remain alive for several hours (even twenty-four hours) before death occurs, depending on the lethal dose of insecticide that is has absorbed. Pyrethrum is relatively unstable and loses strength on exposure to air or heat.

Pyrethrum Synergists: An insecticidal synergist is a chemical which, when combined with an insecticide, increases the killing power of the combination beyond what might be expected by the simple addition of each. Common pyrethrum synergists and piperonyl butoxide, sulfoxide, and MFK-264.

Golden Malrin: This effective fly-killer is a sugar bait.

Residual Fly Sprays:
Diazinon: 5-1%
Malathion: 2%
Rotenone Baygon: 1%

Note: Chlordane, DDT, and Lindane are banned in New York State. Parathion is a dangerous organic phosphate and under no conditions may it be used in home or food establishments.

INSECT CONTROL: ROACHES

The elimination of roaches in food establishments is a greater problem than flies. Roaches also carry disease bacteria on their bodies and deposit them on the food through their excreta, vomit, and bodily contact.

The most common household roach is the German roach. It is one-half to one inch in size, light brown in color with two dark stripes on the back, very agile, and very prolific. The female lays twenty-five to thirty eggs at one time, enclosed in a leathery pouch one-quarter inch long. The life span of a roach is one to one and a half years, and the female will lay five batches of eggs in one year. The egg hatches into a nymph and the nymph grows by molting its shell-like skin. After five molts, the roach reaches adult size. The length of time required for hatching eggs depends on the temperature, and ranges from one to five months.

Roaches are cunning and quick to sense danger. They can survive for long periods under unfavorable conditions and their flat oval shapes facilitate hiding in cracks and crevices into which they can glide with lightning rapidity. They generally travel and feed at night. While they prefer starchy foods, they can feed on leather, felt, or wood. All loose foods should be kept in covered containers.

CONTROL

I. It is difficult to prevent the invasion of a food establishment with insects, especially roaches, thay may come from an adjoining building or in packages delivered to the plant. The emphasis must be placed on eliminating of harborages and breeding places within the establishment as well as extermination.

II. All cracks and holes in the floor, walls, and ceilings should be eliminated as far as possible by filling with cement, plaster, putty, or plastic wood. Seams in fixtures and equipment should receive the same treatment. III. Equipment and fixtures should be placed flush against the wall and floor; if not, then a sufficient distance away from the wall and above the floor to facilitate cleaning around it. Wherever possible, wooden fixtures should be replaced with metal.

IV. All potential insect breeding places, such as rubbish, debris and stagnant water, should be eliminated. Garbage should be kept in tightly covered metal cans, and the cans should be thoroughtly cleaned after being emptied. The room in which garbage is kept, prior to removal, should be constructed of impervious washable material, preferably cement, and should have can washing facilities. If this room can be refrigerated, the cold temperature will prevent insects from breeding, and odors from decomposing garbage will be inhibited.

V. Good housekeeping is a very important factor in insect control. The food establishment and equipment therein should be completely cleaned each night before closing, not only for good sanitation, but to remove all grease, food encrustation, and food particles on which the insects can feed.

VI. In addition, roaches can be destroyed with effective insecticides.

DESTRUCTION BY POISON (EXTERMINATION)

I. Space sprays are not very effective against roaches. The spray or powder used must actually contact the body and so aerosol types of sprays are not used for roaches. Where powder is used, it is generally spread with a mechanical device or “gun”.

II. Insecticides may be classified as contact poisons or stomach poisons. A contact poison is one that only requires contact with the preparation to injure or destroy the insect. This occurs when the insecticide enters the external breathing apparatus or is deposited upon other vulnerable portions of the body. A stomach poison is one which is required to enter the insect's digestive apparatus in order to kill the pest. This may occur when poison is mixed with the food bait so that it is eaten by the insect, or when the insect walks through an insecticide (which may be a contact or stomach poison) and grooms itself, (licks its feet and antennae,) inadvertently ingesting the poison.

III. In extermination by insecticides, it is important to remember that the Health Code prohibits the use of any poisonous insecticides in a food establishment. Artificially colored blue fluoride powder is the only exception to this rule. When new insecticides are offered on the market for use in food establishments, they must be accepted by the Fumigant Board of the Health Department as to their non-poisonous character.

Insecticides commonly used for roach destruction are:

I. Pyrethrum: Non-poisonous to man but effective against roaches. It is non-stable and loses its strength rapidly when exposed to air, moisture, and sun. It is used in several forms, usually as a powder or spray. It is also used in the form of a pyrethrum vapor that penetrates every crack and crevice in the room. Various devices exist in which steam is generated that passes through a narrow horizontal nozzle to which is attached a bottle of pyrethrum dissolved in kerosene. The stream of steam causes this solution of pyrethrum to vaporize throughout the room in which the device is used, and the roaches leave their hiding places to get air and generally die on the floor where they can be swept up.

II. Rotenone: Is non-poisonous to man. It is a product extracted from plants known as Derris Root, Cube, and Timbo. It is an effective roach killer and is generally spread as a powder, although it may also be used in sprays.

SUSPECTED ROACH INFESTATION SUGGESTIONS FOR INSPECTION

1. Locations where commonly found Kitchen, vegetable preparation area, serving counter, garbage area, walls and shelving adjacent thereto; behind hanging pictures and signs, beneath paper shelf lining, at electric and steam meters. All warm, slightly moist, seldom disturbed dark spots.

2. Structural Defects Harborages as cracks, crevices, seams, fissures in walls, fixtures, furniture, pipes, and wire passageways.

3. Storage Conditions

   A. Accumulations of seldom disturbed paper goods as menus, newspapers, paper stock, bags, cups, cartons, containers, straws.
   B. Accumulations of empty unrinsed food containers as milk, beer and soda bottles, ice cream containers, milk cans, meat, fish and bakery boxes.

4. Insanitary Conditions

   A. Unclean garbage can exterior surface and adjacent floor area.
   B. Dried food scraps on surfaces of floor, walls, shelving; beneath and adherent to fixtures and equipment such as mixers, slicers, dicers, peeler impact about can opener and within seams of cutting boards, chopping blocks and work tables.

PESTICIDE USE IN FOOD SERVICE OPERATIONS

ARE YOU HARMING.....

.....yourself?
.....your employees?
.....your customers?

You may be - without realizing it - if you are using pestices improperly.
You may also be breaking the law.

THE LAW

State law requires that anyone using pesticides, except farmers or homeowners on their own property, be certified as a "commercial applicator" with the State Department of Environmental Conservation (DEC). This applies even to arerosol cans of pesticides, available at most grocery stores.

Persons applying pesticides in restaurants, institutional kitchens or other food service operations need special certification from DEC's Bureau of Pesticides Management. Under the State Sanitary Code, a food service operation can be closed down by the State Health Department if pesticides are applied by a person not certified to do so, or if pesticides or other toxic substances are improperly used, stored or labeled.

WHY?

All pesticides are toxic. After all, their purpose is to kill roaches, rats and other pest. Proper use of pesticides is essential to everyone's safety.

If pesticides are used improperly, food, utensils or other food-preparation equipment can become contaminated. Improper handling of pesticides can also lead to direct exposure through the nose, mouth or skin. The result may be future health problems for those exposed.

State or local health department sanitarians check for proper labeling, storage and use of pesticides during their routine inspections of food service establishments. Sanitarians also make sure that only registered businesses or certified persons are applying pesticides. Violations found during these inspections can result in fines, closure or both.

DO'S AND DON'TS

Help insure the safest control of pests in your food service operation by following these do's and don'ts:

DO be concerned about your health and safety, and that of your employees and customers.

DO ask your pest controller for proof of certification by DEC'S Bureau of Pesticides Management.

DO report uncertified pest controllers to a DEC pesticide inspector at the DEC pesticide inspector at the DEC Regional Office.

DO make a pest's life difficult by maintaining extra-clean conditions and by eliminating possible pest entry routes.

DO know what pest species you are dealing with, and what pesticides are being used. Ask your pest controller or health department sanitarian about options for pest control, including chemical and non-chemical methods. (A combination of both is often most effective.)

DO get instructions from both the pest controller and a sanitarian on what you should do following treatment.

DO read and follow the label instructions carefully when using any pesticide product.

DON'T hire an uncertified pest controller.

DON'T apply any pesticide in a food service operation yourself unless you are certified.

DON'T permit the application of pesticides while food is being prepared or served, or in an area where utensils, unprotected food or containers are stored.

FOR MORE INFORMATION .....

To learn more about pest control, or for details on how to become certified to apply pesticides, contact your local health department or DEC regional pesticide office.

RODENT AND INSECT CONTROL

HOW CAN WE CONTROL RATS AND MICE?

I. GET RID OF THEIR NESTING PLACES
Clean up all piles of rubbish, inside and outside the premises.

II. BUILD THEM OUT
Block all possible rat entrances. Rat-proof foundations.

III. STARVE THEM OUT
Protect food at night. Keep garbage containers closed. Do a thorough clean-up job.

IV. KILL THEM
Use traps for temporary control.

HOW CAN WE CONTROL FLIES?

I. GET RID OF THEIR BREEDING PLACES
Control the sources.

II. KEEP THEM OUT
Screen doors and windows properly. See that all doors open out and are self-closing. Install overhead fly fans or air curtains.

III. DO A GOOD JOB OF HOUSEKEEPING
Keep foods covered. Keep garbage containers sealed. Remove food accumulations promptly.

IV. KILL THEM
Use a pyrethrin insect spray inside the buildings.CAUTION: Don't use sprays with any food or food surfaces exposed in the room.

HOW CAN WE CONTROL COCKROACHES AND OTHER INSECTS?

I. BE ALERT TO FIRST SIGNS OF INFESTATION
Destroy infested foods.

II. DO A GOOD JOB OF HOUSEKEEPING AND STORAGE

III. USE PROPER INSECTICIDES (CAREFULLY)

THEN IF RATS, MICE, FLIES, COCKROACHES OR OTHER INSECTS STILL INFEST YOUR ESTABLISHMENT, CONTACT YOUR LOCAL HEALTH DEPARTMENT!

BASIC FUNDAMENTALS OF INSECT AND RODENT CONTROL

CONTENTS

BASIC FUNDAMENTALS OF INSECT AND RODENT CONTROL

PURPOSE: Prevent entry, reduce the threat of infection and disease and to provide a sanitary environment.

EQUIPMENT:

Knowledge of life cycle, habitation and signs of infestations
Good housekeeping
Denial of entry and access to food
Chemical insecticides guided by the Environmental Protection Agency (EPA)

SAFETY PRECAUTIONS:

1. Be aware: Insects and rodents are not just nuisance to man, they are regarded as health hazards. Insects carry disease and bacteria and rodents carry fleas that transmit Typhus and Bubonic Plague.

2. Report signs of insects or rodents to Sanitary Engineering Branch.

3. Be safety conscious of chemical insecticides. Use poisons and insecticides under supervision only, and as directed by EPA.

4. Keep chemical poisons away from food, covered, clearly labeled and stored in locked cabinets.

5. Never use insecticides before reading label.

6. Make sure insecticide is registered by the U.S.D.A. before purchasing.

1. BRIEF HISTORY:

There are over 500,000 species of insects but only 1% are of any real concern to man. This 1% causes 200 different types of diseases. Their sizes vary from $^{1}/_{16}$ of an inch to 7 inches. Among the insects causing the most disease and destruction are: cock-roaches, flies, mosquitoes, ants, bed bugs, moths, lice, termites and fleas.

2. COCKROACHES:

Cockroaches are pale tan, mahogany or black in color with long slender antennae.

The 4 types of roaches most commonly found in institutions are:

1. The German roach—Blatella Germanica—the craton bug.
2. Brown-banded roach—Supella Supelectillium.
3. Oriental roach—Blatella Orientalis—Shad roach.
4. American Roach—Periplanetta Americana—The Bombay Canary.

The Cockroach is an ancient creature that has adapted itself to the abode of man and can be the most troublesome insect in an institution. The offensive secretions of these insects, and their loathsome habit of crawling on filth, then onto dishes and food, has made them among the most despised of vermin. They are also potential vectors of disease.

Control measures include complete cleanliness and removal or denial of access of food, storing foods properly in closed containers and use of insecticides. Some authorities state that the powder insecticide is the most effective control for roaches. However, roaches have built up a resistance to insecticides.

REPORTS OF VIRGINIA POLYTECHNICAL INSTITUTE (AS OF DECEMBER 1960) FOR RESISTANT COCKROACHES.

Today more than ever in the history of the world, man is becoming acutely aware of the stability of insect life, for man with all of his knowledge and wisdom, his scientific research, and modern laboratories realizes that he is dealing with nature in all of its mystic powers. No matter what he develops in the way of control methods, nature has always had the ability to overthrow its potentialities.

Almost every person today is confronted with some insect problem, whether it be roaches, flies, fleas, ticks or moths. As he investigates the market for materials to combat these insects, he finds that he gets very little results and because of this, soon realizes that something strange is going on in the insect world.

What is happening? This is the problem that our scientists are trying to solve today, the question of just how and why these insects are able to build up immunity against such powerful chemicals compounded out of hydrocarbons and phosphates.

In 1947, studies were begun at V.P.I. by Dr. J. M. Grayson, who was joined in 1957, by Dr. D. G. Cochron using a selection of populations of German roaches for resistance to various insecticides. There are now six or seven different projects being carried out in the insecticide resistance areas.

For the last 12 to 14 years the standard control procedure for German roaches has been 2% solution of Chlordane. It is now known that Chlordane resistance is common in áll parts of the country, where this has happened the insecticide Diazinon is recommended, however, there has been reports that resistance to Diazinon is also developing.

The problem of stability or how long will resistance hold up if the insects are no longer exposed to the insecticides, is one which these scientists have been trying to determine. There had not been enough time to determine the longevity of the phosphate insecticides; however, they do have sufficient information concerning the hydrocarbon insecticides such as Chlordane, D.D.T., Lindane. They have found that resistance to D.D.T. is lost quite rapidly, and when the insects are re-exposed, resistance develops very slowly; however, with Chlordane and Lindane, resistance lasts for a considerably longer time. With resistance of 20% after 25 generations of insects that are not exposed to any chemicals and when re-exposed, these insects build resistance in two generations. So, we find that the problem of resistance to insecticides by the German Roach is one which warrants continuing studies.

Dr. Grayson carried out extensive studies, testing new materials and formulating new materials against resistant roaches. The following are a few of the chemicals tested and their results:

1. The compound Chemagro 29493 or Baytex as it is called gave good results in a 1% emulsion form.

2. The results in oil were not as good.

3. The material Butonate at the .8% level in oil

was not very promising; however, at a higher percentage it may do a better job.

4. The material D.D.V.P. did not hold up over a 60-day period in a 2% solution with Dieldrin. This material however, at a .5% might prove effective and this percentage has recently been approved by the pesticide regulation section of the U.S.D.A. for limited use by P.C.O. only. This action is based upon the fact that D.D.V.P. in the vapor form is not as hazardous as previously thought. It is still, however, a dangerous material, and should be used with a great deal of care.

5. The material, Hercules 7522C, which is a carbonate material was *not* effective in these tests.

6. Malathion SF60, which is a formulation of Malathion designed to give longer residual action was proven not to be good, however, it did work in some areas.

7. The compound, Monsanto CP 11223 is an excellent roach control material; however, because of its limitations as far as other insects are concerned the manufacturer feels that the roach market is not sufficient enough to warrant the manufacturing of this material. We are given to think that this material will not be pursued further.

8. The compound Diazinon in a 5% solution was found to be the standard in these tests and gave good results.

3. FLIES:

A winged insect that emerges from larvae—or maggots, as they are most commonly called and feed on the materials of their surroundings. There are several species:

1. House Fly—Musca Domesticus.
2. Blow—Phornia Regina.
3. Bottle—Family Calliphoridae.
4. Fruit—Drosophila.
5. The Lesser House Fly—Eannia Canicularis.

The House fly, however, out numbers the other species listed, therefore, the following discussion will pertain to the House fly. It lays its eggs in human and animal excrement. Under favorable (warm) weather conditions, the larvae to adult stage takes 8 to 12 days. Aside from being one of the most annoying insects, it spreads disease.

The primary control measure against flies must be taken at its breeding place. Complete cleanliness and removal of food are the best control against flies. Insecticides are used under supervision only and as directed by EPA.

4. ANTS:

Any of an order of Hymenoptera which includes the bee, the wasp and other related forms that often associate in large colonies with complex social organization. Ants are just nuisance, they have not been implicated as disease spreader. However, they must be controlled in buildings, because of their ability to destroy food and because they are unwelcomed guests. Good sanitation inside buildings and on surrounding grounds is essential.

5. TERMITES:

Any of numerous pale-colored soft-bodied social insects (order of Isoptera) that live in colonies consisting of winged sexual forms, wingless sterile workers and often soldiers. They feed on wood. They have an appearance very *similar* to ants.

Signs of infestation: Clay or dirt packed on basement floor joist (forms the termite tunnels) or pitting of the floor joist.

Call the Sanitary Engineering Branch if any of

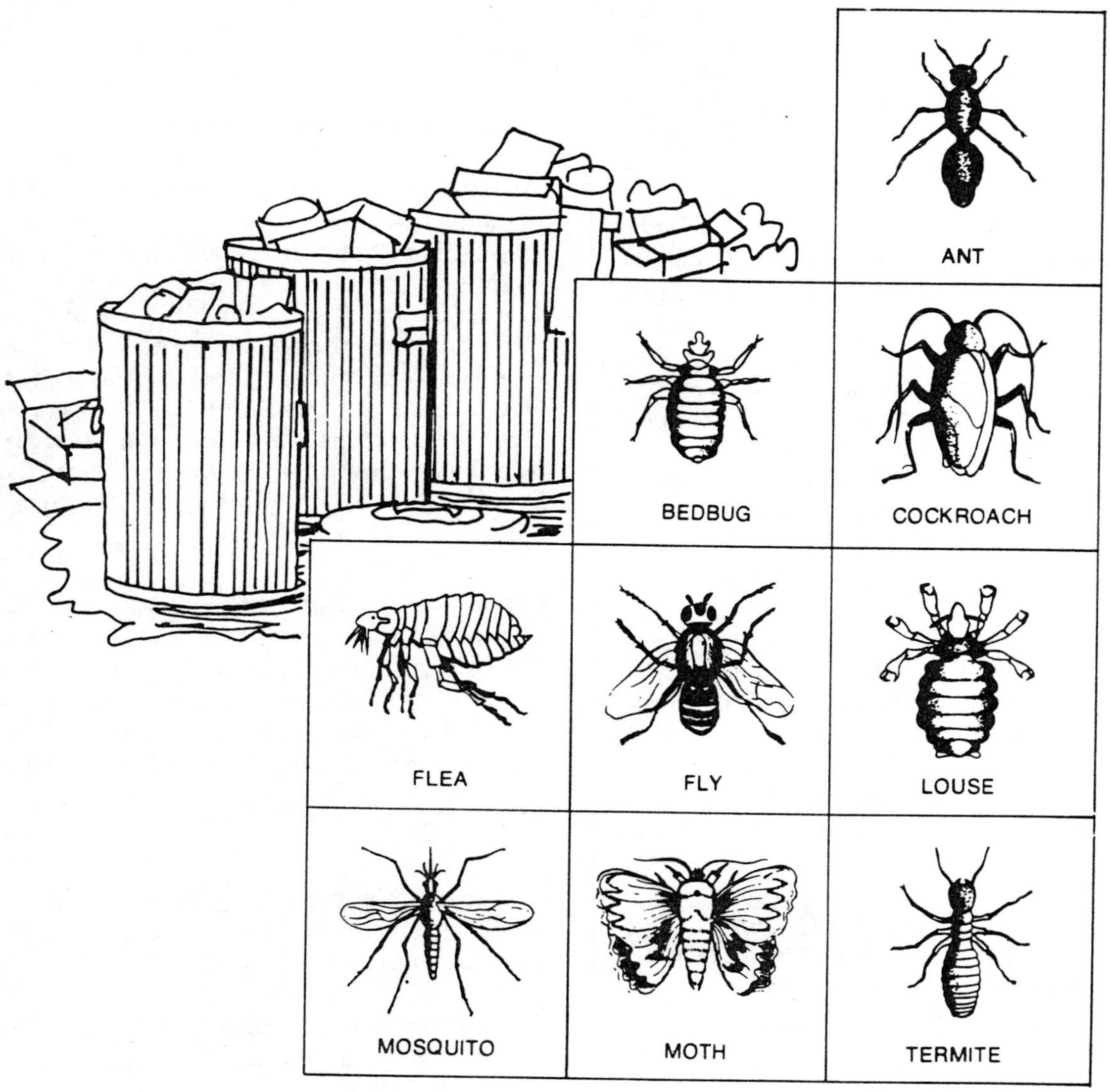

these signs are observed—only professional control measures are effective.

The following are some suggestions that will keep *new* buildings free from these subterranean insects for many years:

1. Site Sanitation: All tree stumps, roots and surplus wood should be removed from the site before construction is started. All surplus wood and grade stakes should be removed as soon as construction is finished. When these items are left in the soil, they become food to establish colonies in the soil.

2. Foundation Structure: It is important to make the foundation impervious to termite attack. Poured reinforced concrete with expansion joints properly filled, makes the best resistant contruction. All wood work resting on, not through the foundation, should be properly treated to protect it from damage. A popular method is the use of Pentachlorophenol, a wood preservative, in oil, soaked into the wood. This type of material after it is dried, permits painting. It is generally used at 5% solution in oil.

3. Ventilation and Drainage: In construction where there are crawl areas, cross ventilation is highly desirable to control the moisture conductive to termite existence. The amount of ventilators needed depends on the area, the humidity and air currents. This is also true of the front steps adjoining the building.

4. Drainage: The soil adjacent to the building should be so graded as to permit the water to drain away from the structure. Proper guttering, draining well away from the corners are vital. All areas under the building must be properly drained.

5. Clearance Beneath the Building: In order to make periodic inspections under the building, a clearance of 18" from the joist to the ground is desirable to facilitate the inspection.

6. Skirting: When used, a clearance of 3 to 6" between it and the ground should be allowed. If this is filled in during the winter, it should be reopened in the spring.

7. Miscellaneous Appendaged: Porches, steps, terraces and platforms should be treated and installed in such a way as to prevent soil from coming into contact with them. Sleepers and studs should not be imbedded in concrete. Plumbing should be installed clear of the ground and not supported by wooden or any other braces that come in contact with the soil. This would permit termites to build tubes that would reach the floor joists. All areas where pipes or conduits enter into the floor or wall from below ground, including expansion joints, should be thoroughly filled with old style roofing pitch.

8. Chemical Soil Barriers: Pretreatment to the soil with residual insecticides should be added to the soil during construction at the rate of 1 gallon of 1% Chlordane emulsion to 10 square feet of flat ground or 2 gallons per linear foot of trenching and footing and wall void treatment.

6. BEDBUGS:

A wingless blood sucking bug—reddish brown in color (Cimex Lectularius). Found in bedding, cracks, crevices of walls and ground baseboards.

Their presence are detected by their odor, blood stains and excretory spots. General cleanliness is the first line of defense.

7. LICE:

The orders, Anoplura and Mallophaga are known collectively as lice. All known numbers of both orders are wingless parasites of warm blooded animals but the members of the two orders differ considerable in the structure and feeding habits. The order Anoplura is made up of two members which have sucking mouthparts and which feed upon blood. The members of the order Mallophaga have chewing mouthparts and feed on skin scales and secretions.

Because all the lice are parasitic and spend virtually all their lifetime on the host, the pest control operator is seldom directly involved in the actual process of louse control. In practically every situation encountered, successful louse control involves direct treatment to the host animal. The responsibility for such treatments belong to physicians and veterinarians.

However, special precautionary measures should be observed as specified when entering the infected area or when handling the belongings of an infested person (observe isolation precaution measures).

Sucking Lice, Order Anoplura: This order is a rather small one, containing approximately 500 species, two of which are parasites of man. The antennae have no more than five segments, the head is narrower than the thorax, and the thoracic segments are fused and cannot be moved independently from the abdomen. In practically all species in which the life history had been studied, the eggs, young and adults are all found on the host. Only mammals are parasitized by this order. The adult female attaches its eggs to the hair of the host, the young and the adults suck blood from the host.

Human Louse Pediculus Humanus SSP. Family: Pediculidae: This species has been the subject of spirited controversy among experts on the classification of this group. Biologically, there are two rather distinct patterns of behavior within the species and many authorities feel that this is actually a complex of two species. These two entities will be considered subspecies in this discussion.

The human louse is important from the standpoint of its activities as a parasite but is even more detrimental to human welfare as a means of spreading the casual organisms of typhus and relapsing fever. Circumstances that cause large numbers of people to be crowded together under unsanitary conditions are most favorable to the rapid development of large louse populations and this fact accounts for the ever present danger of typhus outbreaks in such situations.

Body Louse: Pediculus Humanus Humanus: This is the larger of the two subspecies involved, being generally more than three millimeters long and the integument is relatively soft.

Probably the most important characteristic which differs in the subspecies, is the tendency of the boy louse to remain on the body of the host only during the process of actual feeding. When no feeding, the body louse conceals itself in the clothing of the host. Associated with this behavior is the habit of placing the eggs on the fibers of the garments of the host instead of attaching them directly to the body hair.

This type of egg placement is probably a unique habit which does not occur in other sucking lice.

Since treatment of the clothing of people infested with body lice is an important part of any control measure, the PCO may often play an important part in louse control program. Infested clothing can be sprayed with 0.25% Lindane, 4% Malathion dust.

Head Louse: Pediculus Humanus Capitus: This typical head louse rarely exceeds a length of three millimeters, and the body wall is, in general, a little more deeply pigmented and somewhat tougher than that of the body louse. All stages of the life cycle are spent on the host, the eggs are attached directly to the hairs of the host. Control of this species involves direct treatment of the host and is usually considered a medical problem.

Crab Louse or Pubic Louse: Phthirus Pubis: Family Phthiridae: The crab louse is a parasite of man, and while usually concerned with the infestations of the pubic and perianal regions, it may be found in any region of body hair, all life stages are spent on the body of the host, the eggs are attached directly to the body hairs of the host. This insect does not move around much on the body of the host and may remain stationary with the mouthparts inserted into the skin for several days at a time.

The crab louse is smaller than the body louse and the head louse. Specimens are rarely more than two millimeters long. The body is broad and short and the front legs are much smaller than the second and third pairs of legs. The first apparent abdominal segment bears a total of six spiracles on its upper surface.

Control measures of this insect involves the treatment of the host.

8. RODENTS:

Rats are rodents (Rattus and related genera) that differs from mice by size and features of the teeth. Rats and mice are among the most destructive pests. They destroy food and property; they cause fires and they spread disease (typhus fever, plaque, dysentery, rat-bite fever and others).

There are several types of rats. Those most common in the United States are the Roof Rat (Rattus Rattus) and the Common Rat or Norway Rat (Rattus Norvegicus). The common rat and the house mouse (Mus Musculus) are gnawing animals. Their chisel-like teeth grow constantly in a semi-circular form. They must gnaw to keep these teeth worn down. The common rat is called the Norway Rat, Brown Rat, Barn Rat, Gray Rat and other aliases. They are very prevalent. Conservative estimates place their numbers far above the human population.

The rat is a vicious enemy of man. Therefore, proper controls are necessary. Good rat control measures can be accomplished only if all of the basic control measures are carried out in the following order:

1. Survey
2. Rat proof
3. Eliminate shelter
4. Remove food and water
5. Poison, trap or have them fumigated

IDENTIFICATION AND CONTROL OF INSECTS

CONTENTS

IDENTIFICATION AND CONTROL OF INSECTS

The Insect	*How to Identify*	*How to Control*
ANTS	A tiny insect, either red, yellow or black, it is pinched in at the waist	Find the ants' nest, if possible. When the ants are in the house, note the path which they follow either into or through the house. Discover whether the nests are inside or outside the house. Spray the ants' nests and paths with insecticide. Follow directions.
BED BUGS	An unfed bedbug is brown, about 1/4-inch long and its top has a crinkly look; has no wings; hides during the day, feeds on human blood at night which changes its color to red. Spread by clothing, baggage, old furniture.	Find hiding places and spray them. Bugs leave black or brown spots to mark hideouts. They are often found around seams and folds of mattresses cracks and crevices of bed-steads; behind baseboards - anything not air-tight. Follow directions.
CLOTHES MOTHS AND CARPET BEETLES	Fabric pests, they feed on animal fibers. Moth larvae are 1/2" long, hairless, white with dark heads. Moths are yellowish with about 1/2 inch wingspread. Carpet beetle larvae are 1/4-inch long, long-oval shape, have dark bristles giving them a fuzzy look. Beetles are small and mottled with white, yellow, brown or black; also a black species whose larvae have tapered golden or brown bodies with long brown hair.	Both lay their eggs in clothing, upholstery, concealed places. Larvae feed on wool and other animal fibers - that is, clothing an household furnishings. Adults stay in concealed places such as cracks or drawers, although carpet beetles like sunlight. Control is strictly a matter of cleanliness in every nook of the house. Kill insects living in home by surface spraying. Tight cedar chests keep these pests out of your clothes. Use vacuum cleaner attachments often - to remove lint and hair from hard to get at places. Follow directions.
CRICKETS	Not normally household pests- but they do come in. When they do, they chew holes in fabrics. They leap, have long legs, wings.	In house, close all openings and tighten screens. Apply insecticide around baseboards, in closets, cracks or other places where they may hide. Field crickets develop outdoors and feed on vegetati Enter late in summer. Follow directions.

The Insect	*How to Identify*	*How to Control*
FLEAS	The dog and cat flea are most common and look alike. The human flea is a severe pest in many areas, may live on animals - often on hogs. All fleas require blood of animals or birds to reproduce.	Females of most species lay eggs while they are on animals or birds. The eggs drop to ground, floor, around bedding - and there they hatch. Thus, control of fleas on animals is necessary to prevent infestations in homes. Different insecticides are used for different types of fleas, Follow directions.
FLIES, GNATS, MOSQUITOES, WASPS	We all know what they look like. They are a danger and a menace to health of man and animals. They spread disease germs that may be in the materials they breed in, feed on or walk on.	Breed around manure, garbage and fermentation. Cleanliness a must! If disease germs are in these materials, the insects will spread them. Eliminate breeding places. Cover garbage and empty often. Eliminate all places where water may stand. Use screens and surface sprays. Follow directions but work on prevention.
ROACHES and WATERBUGS	A crawling insect with four legs, hard crusty back. From 1/2" to 2" long, can have yellow stripe on wing or two cross bands of yellow on back. Some light brown with two dark stripes; others are all black or brown.	They like damp places and are run-abouts. Even if your home is free of them - always look for stowaways in delivery boxes. They spread from house to house, are attracted by food. Cover garbage and empty often. Cleanliness is important. Use insecticides around cracks in woodwork, baseboards, waterpipes. Follow directions but work on prevention.
SILVERFISH	A slender, wingless insect covered with scales. About 1/2" long, they are a shiny pale gray; have two feelers at the head and three long tail-like appendages at hind end. More active at night and hide during day.	They thrive best in damp, cold places, especially basements, near laundries. Their greatest damage is done eating foods and other materials high in protein, starch or sugar. Eat cereals, moist wheat flour, glued or bound paper. Spray insecticide around the baseboards, doors, windows, and pipes. Follow Directions.

The Insect	*How to Identify*	*How to Control*
SPIDERS	Several kinds get into the house. You will know you have a spider if you see the webs in corners, on furniture and in other places.	Spiders feed on flies an other insects. (But it no cure for flies - to have spiders !) They enter homes through scre or around loose-fitting windows or doors. Tight screens are necessary. flies and other insects are kept out, spiders ar not encouraged to enter. Breed in trash, piles of lumber, and weeds. Follow directions.
TICKS	The Brown Dog Tick is most common of home ticks. More apt to infest homes if dogs are present. A hard shield on their back is marked with a small white dot or two small crescents.	Attach themselves to dog May be found in cracks, hind cartons, under furn ture, rugs, baseboards o moldings. Heated buildi reduces survival. May transmit fevers. Surfac spray. Follow direction

ANSWER SHEET

EST NO. ______________ PART ______ TITLE OF POSITION ______________________________
(AS GIVEN IN EXAMINATION ANNOUNCEMENT - INCLUDE OPTION, IF ANY)

PLACE OF EXAMINATION ______________________ (CITY OR TOWN) ______________________ (STATE) DATE ____________

RATING

USE THE SPECIAL PENCIL. MAKE GLOSSY BLACK MARKS.

1	A	B	C	D	E	26	A	B	C	D	E	51	A	B	C	D	E	76	A	B	C	D	E	101	A	B	C	D	E
2	A	B	C	D	E	27	A	B	C	D	E	52	A	B	C	D	E	77	A	B	C	D	E	102	A	B	C	D	E
3	A	B	C	D	E	28	A	B	C	D	E	53	A	B	C	D	E	78	A	B	C	D	E	103	A	B	C	D	E
4	A	B	C	D	E	29	A	B	C	D	E	54	A	B	C	D	E	79	A	B	C	D	E	104	A	B	C	D	E
5	A	B	C	D	E	30	A	B	C	D	E	55	A	B	C	D	E	80	A	B	C	D	E	105	A	B	C	D	E
6	A	B	C	D	E	31	A	B	C	D	E	56	A	B	C	D	E	81	A	B	C	D	E	106	A	B	C	D	E
7	A	B	C	D	E	32	A	B	C	D	E	57	A	B	C	D	E	82	A	B	C	D	E	107	A	B	C	D	E
8	A	B	C	D	E	33	A	B	C	D	E	58	A	B	C	D	E	83	A	B	C	D	E	108	A	B	C	D	E
9	A	B	C	D	E	34	A	B	C	D	E	59	A	B	C	D	E	84	A	B	C	D	E	109	A	B	C	D	E
10	A	B	C	D	E	35	A	B	C	D	E	60	A	B	C	D	E	85	A	B	C	D	E	110	A	B	C	D	E

Make only ONE mark for each answer. Additional and stray marks may be counted as mistakes. In making corrections, erase errors COMPLETELY.

11	A	B	C	D	E	36	A	B	C	D	E	61	A	B	C	D	E	86	A	B	C	D	E	111	A	B	C	D	E
12	A	B	C	D	E	37	A	B	C	D	E	62	A	B	C	D	E	87	A	B	C	D	E	112	A	B	C	D	E
13	A	B	C	D	E	38	A	B	C	D	E	63	A	B	C	D	E	88	A	B	C	D	E	113	A	B	C	D	E
14	A	B	C	D	E	39	A	B	C	D	E	64	A	B	C	D	E	89	A	B	C	D	E	114	A	B	C	D	E
15	A	B	C	D	E	40	A	B	C	D	E	65	A	B	C	D	E	90	A	B	C	D	E	115	A	B	C	D	E
16	A	B	C	D	E	41	A	B	C	D	E	66	A	B	C	D	E	91	A	B	C	D	E	116	A	B	C	D	E
17	A	B	C	D	E	42	A	B	C	D	E	67	A	B	C	D	E	92	A	B	C	D	E	117	A	B	C	D	E
18	A	B	C	D	E	43	A	B	C	D	E	68	A	B	C	D	E	93	A	B	C	D	E	118	A	B	C	D	E
19	A	B	C	D	E	44	A	B	C	D	E	69	A	B	C	D	E	94	A	B	C	D	E	119	A	B	C	D	E
20	A	B	C	D	E	45	A	B	C	D	E	70	A	B	C	D	E	95	A	B	C	D	E	120	A	B	C	D	E
21	A	B	C	D	E	46	A	B	C	D	E	71	A	B	C	D	E	96	A	B	C	D	E	121	A	B	C	D	E
22	A	B	C	D	E	47	A	B	C	D	E	72	A	B	C	D	E	97	A	B	C	D	E	122	A	B	C	D	E
23	A	B	C	D	E	48	A	B	C	D	E	73	A	B	C	D	E	98	A	B	C	D	E	123	A	B	C	D	E
24	A	B	C	D	E	49	A	B	C	D	E	74	A	B	C	D	E	99	A	B	C	D	E	124	A	B	C	D	E
25	A	B	C	D	E	50	A	B	C	D	E	75	A	B	C	D	E	100	A	B	C	D	E	125	A	B	C	D	E

ANSWER SHEET

TEST NO. __________ PART ______ TITLE OF POSITION ______________________________
(AS GIVEN IN EXAMINATION ANNOUNCEMENT - INCLUDE OPTION, IF ANY)

PLACE OF EXAMINATION ______________________________ DATE __________
(CITY OR TOWN) (STATE)

RATING

USE THE SPECIAL PENCIL. MAKE GLOSSY BLACK MARKS.

1 A B C D E	26 A B C D E	51 A B C D E	76 A B C D E	101 A B C D E
2 A B C D E	27 A B C D E	52 A B C D E	77 A B C D E	102 A B C D E
3 A B C D E	28 A B C D E	53 A B C D E	78 A B C D E	103 A B C D E
4 A B C D E	29 A B C D E	54 A B C D E	79 A B C D E	104 A B C D E
5 A B C D E	30 A B C D E	55 A B C D E	80 A B C D E	105 A B C D E
6 A B C D E	31 A B C D E	56 A B C D E	81 A B C D E	106 A B C D E
7 A B C D E	32 A B C D E	57 A B C D E	82 A B C D E	107 A B C D E
8 A B C D E	33 A B C D E	58 A B C D E	83 A B C D E	108 A B C D E
9 A B C D E	34 A B C D E	59 A B C D E	84 A B C D E	109 A B C D E
10 A B C D E	35 A B C D E	60 A B C D E	85 A B C D E	110 A B C D E

Make only ONE mark for each answer. Additional and stray marks may be counted as mistakes. In making corrections, erase errors COMPLETELY.

11 A B C D E	36 A B C D E	61 A B C D E	86 A B C D E	111 A B C D E
12 A B C D E	37 A B C D E	62 A B C D E	87 A B C D E	112 A B C D E
13 A B C D E	38 A B C D E	63 A B C D E	88 A B C D E	113 A B C D E
14 A B C D E	39 A B C D E	64 A B C D E	89 A B C D E	114 A B C D E
15 A B C D E	40 A B C D E	65 A B C D E	90 A B C D E	115 A B C D E
16 A B C D E	41 A B C D E	66 A B C D E	91 A B C D E	116 A B C D E
17 A B C D E	42 A B C D E	67 A B C D E	92 A B C D E	117 A B C D E
18 A B C D E	43 A B C D E	68 A B C D E	93 A B C D E	118 A B C D E
19 A B C D E	44 A B C D E	69 A B C D E	94 A B C D E	119 A B C D E
20 A B C D E	45 A B C D E	70 A B C D E	95 A B C D E	120 A B C D E
21 A B C D E	46 A B C D E	71 A B C D E	96 A B C D E	121 A B C D E
22 A B C D E	47 A B C D E	72 A B C D E	97 A B C D E	122 A B C D E
23 A B C D E	48 A B C D E	73 A B C D E	98 A B C D E	123 A B C D E
24 A B C D E	49 A B C D E	74 A B C D E	99 A B C D E	124 A B C D E
25 A B C D E	50 A B C D E	75 A B C D E	100 A B C D E	125 A B C D E